Non-Volatile In-Memory Computing by Spintronics

Synthesis Lectures on Emerging Engineering Technologies

Editor
Kris Iniewski, *Redlen Technologies, Inc.*

Non-Volatile In-Memory Computing by Spintronics
Hao Yu, Leibin Ni, and Yuhao Wang
2016

Layout Techniques for MOSFETs
Salvador Pinillos Gimenez
2016

Resistive Random Access Memory (RRAM)
Shimeng Yu
2016

The Digital Revolution
Bob Merritt
2016

Compound Semiconductor Materials and Devices
Zhaojun Liu, Tongde Huang, Qiang Li, Xing Lu, and Xinbo Zou
2016

New Prospects of Integrating Low Substrate Temperatures with Scaling-Sustained Device
Architectural Innovation
Nabil Shovon Ashraf, Shawon Alam, and Mohaiminul Alam
2016

Advances in Reflectometric Sensing for Industrial Applications
Andrea Cataldo, Egidio De Benedetto, and Giuseppe Cannazza
2016

Sustaining Moore's Law: Uncertainty Leading to a Certainty of IoT Revolution
Apek Mulay
2015

Non-Volatile In-Memory Computing by Spintronics

Hao Yu, Leibin Ni, and Yuhao Wang

ISBN: 978-3-031-00904-4   paperback
ISBN: 978-3-031-02032-2   ebook

DOI 10.1007/978-3-031-02032-2

A Publication in the Springer series
SYNTHESIS LECTURES ON EMERGING ENGINEERING TECHNOLOGIES

Lecture #8
Series Editor: Kris Iniewski, *Redlen Technologies, Inc.*
Series ISSN
Print 2381-1412   Electronic 2381-1439

Non-Volatile In-Memory Computing by Spintronics

Hao Yu
Nanyang Technological University, Singapore

Leibin Ni
Nanyang Technological University, Singapore

Yuhao Wang
Synopsis, California, USA

SYNTHESIS LECTURES ON EMERGING ENGINEERING TECHNOLOGIES #8

ABSTRACT

Exa-scale computing needs to re-examine the existing hardware platform that can support intensive data-oriented computing. Since the main bottleneck is from memory, we aim to develop an energy-efficient in-memory computing platform in this book. First, the models of spin-transfer torque magnetic tunnel junction and racetrack memory are presented. Next, we show that the spintronics could be a candidate for future data-oriented computing for storage, logic, and interconnect. As a result, by utilizing spintronics, in-memory-based computing has been applied for data encryption and machine learning. The implementations of in-memory AES, Simon cipher, as well as interconnect are explained in details. In addition, in-memory-based machine learning and face recognition are also illustrated in this book.

KEYWORDS

Spintronics, in-memory computing, non-volatile memory, logic-memory integration, data encryption, AES, machine learning, data analytics, hardware accelerator

Contents

Preface

The existing memory technologies have critical challenges when scaling at nanoscale due to process variation, leakage current, and I/O access limitations. Recently, there are two research trends attempting to alleviate the memory-wall issues for future big-data storage and processing system.

First, the emerging non-volatile memory technologies, such as spin-transfer torque memory, domain wall nanowire (or racetrack) memory, etc., have shown significantly reduced standby power and increased integration density, as well as close-to DRAM/SRAM access speed. Therefore, they are considered as promising next generation memory for big-data applications.

Second, due to high data-level parallelism in big-data applications, large number of application specific accelerators can be deployed for data processing. However, the I/O bandwidth limitation will still be the bottleneck in such a memory-logic integration approach. Instead, an in-memory computing platform will be highly desired with less dependence on I/Os.

In order to achieve low power and high throughput (or energy efficiency) in big-data computing, one can build an in-memory non-volatile memory (NVM) hardware platform, where both the memory and computing resources are based on NVM devices with instant-switch-on as well as to ultra-low leakage current. Such an architecture can achieve significant power reduction due to the non-volatility. Moreover, one can develop an NVM logic accelerator that can perform domain-specific computations such as data encryption and also data analytics in a logic-in-memory fashion. Compared to conventional memory-logic-integration architectures, the storage data do not need to be loaded into volatile main memory, processed by logic, and written back afterward with significant I/O communication congestion.

In this book, we discuss the following research studies on this regard. First, we introduce a non-volatile in-memory architecture that both data storage and logic computing are inside the memory block. The data storage and logic computing is located in pairs to perform a distributed in-memory computing. We illustrate the NVM-based basic memory and logic components and find significant power reduction. Second, we develop a SPICE-like simulator NVM SPICE, which implements physical models for non-volatile devices in a similar way as the BSIM model for MOSFET. We further develop the data storage and logic computing on spintronic devices. These operations can be deployed for in-memory data encryption and analytics. Moreover, we illustrate how data encryption including advanced encryption standard (AES) and Simon cipher can be implemented in this architecture. Data analytics applications such as machine learning for super-resolution and face recognition are performed on the developed NVM platform as well.

This book provides a state-of-the-art summary for the latest literature on emerging non-volatile spintronic technologies and covers the entire design flow from device, circuit to system perspectives, which is organized into five chapters. Chapter 1 introduces the basics of conventional memory architecture with traditional semiconductor devices as well as the non-volatile in-memory architecture. Chapter 2 details the device characterization for spintronics by non-electrical states and the according storage and logic implementation by spintronics. Chapter 3 explores the in-memory data encryption based on STT-RAM as well as domain wall nanowire. The implementations of in-memory AES, domain wall-based Simon cipher and low power interconnect are explained in detail. Chapter 4 presents the system level architectures with data analytics applications for the emerging non-volatile memory. In-memory machine learning and face recognition are discussed in this chapter. This book assumes that readers have basic knowledge of semiconductor device physics. It will be a good reference for senior undergraduate and graduate students who are performing research on non-volatile memory technologies.

Hao Yu, Leibin Ni, and Yuhao Wang
Singapore
October 2016

Acknowledgments

The authors would like to thank their colleagues at CMOS Emerging Technology Group at Nanyang Technological University: Dr. Wei Fei, Dr. Yang Shang, Dr. Sai Manoj P.D., Mr. Hantao Huang, Mr. Zichuan Liu, Miss. Hang Xu, and Mr. Rai Suleman Khalid. The authors also acknowledge with gratitude discussions with Prof. Guangbin Huang, Prof. Anupam Chattopadhyay, Prof. Chip-hong Chang, Prof. Jianhua Yang, Prof. Kaushik Roy, Prof. Weisheng Zhao, Prof. Dennis Sylvester, Prof. Kevin Cao, Prof. Yuan Xie, Prof. Yiran Chen, Prof. Hai Li, Dr. Tanay Karnik, Dr. Jing Li, Prof. Wei Zhang, Prof. Tony Kim, Prof. Wen-Siang Lew, Prof. Kiat-Seng Yeo, Dr. Rajiv V. Joshi, and Mr. Suman Deb. Their support was invaluable to us during the writing of this book. The relevant research is funded by MOE Tier-2 (MOE2015-T2-2-013) titled "Sparse-represented Non-volatile In-memory Accelerator for Big-Data Analytics" and NRF CRP (NRF-CRP9-2011-01) titled "Magnetic Domain Wall Logic Gate Device," from Singapore.

Hao Yu, Leibin Ni, and Yuhao Wang
October 2016

CHAPTER 1

Introduction

Abstract   Computer memory is any physical device capable to store data temporarily or permanently. It covers from fastest, yet most expensive, static random-access-memory (SRAM) to cheapest, but slowest, hard drive disk, while in between there are many other memory technologies that make trade-offs among cost, speed, and power consumption. However, the large volume of memory will experience significant leakage power, especially at advanced CMOS technology nodes, for holding data in volatile memory for fast accesses. The spin-transfer torque magnetic random-access memory (STT-RAM), a novel non-volatile memory- (NVM) based on spintronic devices, has shown the great benefits on power-wall issue compared to traditional volatile memories. In addition, in traditional Von-Neumann architecture, the memory is separated from the central processing unit. As a result, the I/O congestion between memory and processing unit leads to the memory-wall issue, and ultimate solution requires a breakthrough on memory technology. The novel in-memory architecture is the solution of memory-wall issues that both logic operation and data storage are located inside memory. This chapter reviews the existing semiconductor memory technologies and traditional memory architecture first, and then introduces the spintronic memory technologies as well as the in-memory architecture.

1.1   MEMORY WALL

The internet data has reached exa-scale (10^{18} bytes), which has introduced emerging need to re-examine the existing hardware platform that can support intensive data-oriented computing. At the same time, the analysis of such a huge volume of data needs a scalable hardware solution for both big-data storage and processing, which is beyond the capability from pure software-based data analytic solution. The main bottleneck is from the well-known memory bottleneck. As such, one needs to re-design an energy-efficient hardware platform for future big-data driven application.

A big-data-driven application requires high-bandwidth with maintained low-power density. For example, web-searching application involves crawling, comparing, ranking, and paging of billions of web-pages or images with extensive memory access. The microprocessor needs to process the stored data with intensive memory access. However, the present data storage and processing hardware have well-known bandwidth-wall due to limited accessing bandwidth at I/Os; but also power-wall due to large leakage power in advanced CMOS technology when holding data. As such, a design of scalable energy-efficient big-data analytic hardware is highly required.

The following research activities attempted to alleviate the memory bottleneck problem for the future big-data storage and processing.

- Emerging non-volatile in-memory computing is a promising technology for big-data oriented computing. The application-specific accelerators can be developed within memory such that the data will be pre-processed before they are read out with the minimum number of data migrations such that the bandwidth-wall can be relieved. Moreover, the non-volatile memory devices hold information without using charge, such that leakage current can be significantly reduced with the power-wall relieved as well.

- Sparse-represented data by compressive sensing is an effective approach to reduce data size by projecting data from high-dimensional space to low-dimensional subspace with essential feature preserved. Data stored in memory after compressive sensing can be recovered or directly processed. This procedure reduces data complexity in data storage as well as in data analytics with a further significant power reduction and bandwidth improvement.

- Data analytics by a machine-learning accelerator can be developed for fast data analytics. The latest machine learning algorithm on-chip can accelerate the learning process with a potential to realize online training, which is important for applications in autonomous vehicles and unmanned aerial vehicles.

In this book, we aim at finding a scalable hardware solution based on the concept of non-volatile in-memory data-analytics accelerator. In order to achieve low power and high bandwidth (or energy efficient) in big-data-driven computing, we propose developing a non-volatile memory- (NVM) based hardware platform, where both the data storage and processing are based on NVM devices with instant-switch-on as well as ultra-low leakage current. This can result in significant power reduction due to the non-volatility. Moreover, we will develop NVM-based logic accelerator that can perform domain-specific computation such as machine-learning with a logic-in-memory architecture. In contrast for the conventional memory-logic-integration architecture, the stored data must be loaded into volatile main memory, processed by logic, and written back with significant I/O communication and power overhead. Last, we will develop a machine-learning algorithm based on sparse-represented data to further deal with large-scale data analytics. As a result, we outline the following detailed studies in this book.

- First, we plan to develop an energy-efficient in-memory computing architecture by leveraging the non-volatile memory devices that can fundamentally resolve the memory bottleneck for big-data analytics. In this book, non-volatile-memory device such as spin-transfer torque magnetic random-access memory (STT-RAM) or domain wall nanowire (DW-NN) will be studied to realize both data storage and also in-memory logic. As such, it can greatly reduce the leakage power due to its non-volatile state without holding electrons in nature. In addition, the logic-in-memory architecture will also be studied to perform logic computation locally, thus relaxing the demand on I/O bandwidth and alleviating the bandwidth constraint.

- Second, we plan to develop data encryption on NVM-based platform. The Advanced Encryption Standard (AES) and the Simon cipher are specifications for the encryption of electronic data. In this book, the detail mapping of these two encryption approaches on the in-memory computing architecture will be studied. As such, the system throughput can be improved and the area overhead can be reduced. Moreover, the optimizations of hardware implementation such as pipeline structure will be discussed.

- Third, we plan to develop data analytics on the in-memory architecture. Machine learning is the primary data analysis approach that solves problems in a "programming by example" or "learning process" fashion. In this book, we will study how to develop machine-learning accelerators that can: (1) deal with sparse-represented data; and (2) work in a non-volatile in-memory computing architecture. We will show how to realize such a machine-learning algorithm on domain wall nanowire toward super resolution. In addition, sparse-represented data on STT-RAM will be studied for face recognition. By utilizing this architecture, significant energy efficiency can be achieved.

The main contribution of this book is the non-volatile, memory-based, in-memory computing architecture. Data encryption and data analytics will be implemented in this architecture with significant energy efficiency. The study of this research can help accumulate the future big-data computing intellectual property as well as the according applications for developing new social media, cyber-security, and smart city in Singapore.

1.2 TRADITIONAL SEMICONDUCTOR MEMORY

1.2.1 OVERVIEW

Semiconductor memories refer to the silicon-transistors-based physical devices in computer systems that are used to temporarily or permanently store programs or data. In contrast to storage media like hard disks and optical discs, of which the accesses have to follow a predetermined order due to the mechanical drive limitations, semiconductor memories possess the property of random access, which means that the time it takes to access any data is identical regardless of the data location.

According to volatility, the semiconductor memories can be further classified into volatile memory and non-volatile memory. Volatile memory stores data as electrical signals, and loses its data when the device is turned off. Currently, the most commonly used volatile memories are static random-access memory (SRAM) and dynamic random-access memory (DRAM), whose data is indicated by the electrical voltage levels. On the contrary, non-volatile memory is able to retain its data even when chip is turned off, as its data is mostly preserved by non-electrical states. For instance, bits in programmable read-only memory (PROM) are denoted by whether the fuses of individual memory cells are burned. In this section, the principles and operations of volatile SRAM/DRAM and non-volatile flash memory will be briefly introduced (Figure 1.1).

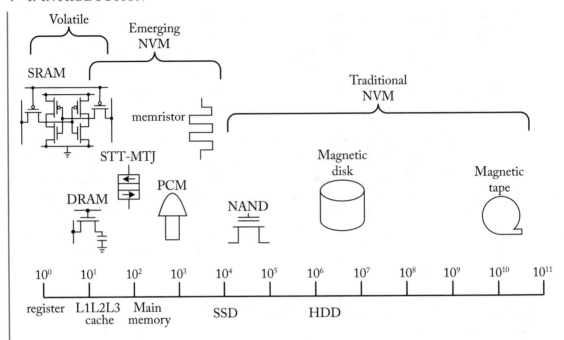

Figure 1.1: Typical memory hierarchy of a computer system (latency measured in processor cycles).

Static Random-access Memory

A typical CMOS static random-access memory (SRAM) cell consists of six transistors, shown in Figure 1.2. The flip-flop formed by M1–M4 holds the stored bit. The term *static* is derived from the fact that the cell does not need to be refreshed like dynamic RAM, and the data can be retained as long as the power is supplied. The M5 and M6, connected with word-line and two bit-lines, are used as access transistors to select target cells.

There are three operation states for a SRAM cell, write, read, and standby. To a write operation, the value to be written needs to be applied on both bit-lines, namely BL and $\overline{BL}$, in a complementary manner. Assume we wish to write "0" to the cell, i.e., Q to be "0" and $\overline{Q}$ to be "1", the BL is driven low and $\overline{BL}$ high. Once the M5 and M6 are turned on by setting WL "1", the bit-line drivers will override the previous stored value. In order to easily override the previous state in the self-reinforced flip-flop, the bit-line drivers are required to be designed stronger than the transistors in flip-flop.

For a read operation, both bit-lines are precharged high before the start of the read cycle, and the turning on of the word-line signifies the start of read operation. Because of the opposite voltages at Q and $\overline{Q}$, one of the bit-lines will be pulled down by the cell, and the discharging of one of the bit-lines is then detected by the bit-line sense amplifier. In voltage mode sensing scheme, the sign of bit-line voltage difference ΔV (V_{BL} minus $V_{\overline{BL}}$) determines the value of stored

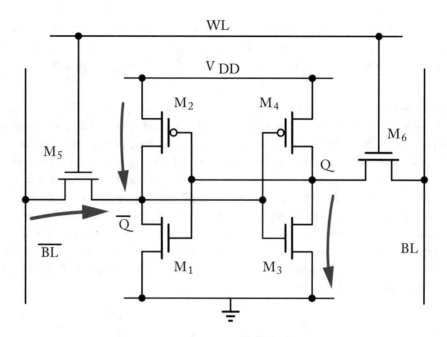

Figure 1.2: A 6T SRAM cell structure with leakage paths in standby state. The bit-lines are precharged high and assume the stored data is "1" at Q.

bit. A ΔV in tens of millivolts is significant enough to efficiently distinguish which bit-line is being discharged. For example, assume the stored bit is "1" at Q and "0" at $\overline{Q}$, and once the WL is asserted, the $\overline{BL}$ will discharge toward "0"; when a positive ΔV of tens of millivolts is gained, the latch-based sense amplifier will amplify the small voltage difference with positive feedback, and finally output logic "1" as result.

When the word-line (WL) is connected to ground, turning off the two access transistors, the cell is in the standby state. During the standby state, the two cross-coupled inverters will reinforce each other due to the positive feedback, the value is preserved as long as the power is supplied. One prominent problem regarding SRAM in standby state is severe subthreshold leakage. Subthreshold leakage is the drain-source current of a transistor when the gate-source voltage is less than the threshold voltage. The subthreshold current depends exponentially on threshold voltage, which results in large subthreshold current in deep sub-micron regime. Figure 1.2 shows three leakage paths in one SRAM cell, assuming the stored bit is "1" at Q. Note that the BL and $\overline{BL}$ are always precharged to V_{DD} to facilitate future read operation. Regardless of the stored value, there always will be three transistors consuming leakage power.

Compared to other memory technologies, SRAM is able to provide the fastest access speed, but the advantage comes as a tradeoff against density and power. As one SRAM cell requires silicon area for six transistors, SRAM has very limited density and hence is more expensive than other memory technologies. In addition, it is very power consuming due to the leakage problem at standby state. Therefore, SRAM serves best in the applications where high performance is the main concern and the capacity is not significant, namely the caches for processors.

Dynamic Random-access Memory
The philosophy behind DRAM is simplicity. Unlike SRAM where one cell is composed of six transistors, each individual DRAM cell consists only one capacitor and one access transistor. The data "0" or "1" is represented by whether the capacitor is fully charged or discharged. However, the electrical charge on the capacitor will gradually leak away, and after a period of time, the voltage on the capacitor is so low for the sense amplifier to differentiate between "1" and "0". Therefore, unlike SRAM that the data can be retained as long as the power is supplied, the retention time for DRAM is finite and all DRAM cells need to be read out and written back periodically to ensure data integrity. Typically, the cells are refreshed once every 32 or 64 ms. This process is known as *refresh*, and this is how the name of dynamic RAM is derived. Figure 1.3 shows the circuit diagram of such 1T1C structure DRAM cell.

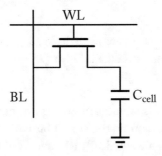

Figure 1.3: The circuit diagram of 1T1C DRAM cell structure.

To write a DRAM cell, the bit-line is first set high or low based on the value to write. After the access transistor is turned on by asserting word-line, the capacitor in the selected cell is charged to "1" or discharged to "0". Because the access takes place through an NMOS transistor, there exists a V_{th} drop during the write "1". In order to prevent this V_{th} drop and maintain a long refresh period, the word-lines driving voltage is usually boosted to $V_{PP} = V_{DD} + V_{th}$.

To read a DRAM cell, the bit-line is precharged to $V_{DD}/2$ and then the word-line is enabled. Due to the charge sharing, the bit-line voltage will slightly decrease or increase depending on the voltage that the capacitor was previously charged to, i.e., V_{DD} or 0. If it is previously charged, the charge sharing will slightly boost the bit-line voltage; otherwise, some charge will be distributed from bit-line to cell capacitor. In both cases, the voltage of storage capacitor will be changed after read operation, thus the read operation is called destructive, and an instant write back is required. A slight voltage change on the bit-line can be calculated by

$$\Delta V = \pm \frac{V_{DD}}{2} \frac{C_{cell}}{C_{cell} + C_{bitline}}. \tag{1.1}$$

The sign of ΔV depends on the state of storage capacitor. In modern DRAM devices, the capacitance of a storage capacitor is far smaller than the capacitance of the bitline. Typically, the capacitance of a storage capacitor is one-tenth of the capacitance of the long bitline that is connected to hundreds or thousands of other cells. The relative capacitance values create the scenario that when the small charge contained in a storage capacitor is placed on the bitline, the resulting voltage on the bitline is small and difficult to measure in an absolute sense. In DRAM devices, the voltage sensing problem is resolved through the use of a differential sense amplifier that compares the voltage of the bitline to a reference voltage.

The use of differential sense amplifier, in turn, introduces some requirements on the DRAM array structure. Particularly, instead of a single bitline, a pair of bitlines needs to be used to sense the voltage value contained in any DRAM cell. In addition, in order to ensure that the voltage and capacitance values on the pair of bitlines are closely matched, the bitlines must be closely matched in terms of path lengths and the number of cells attached. The above requirements lead to two distinctly different array structures: open bitline structures and folded bitline structures.

Flash Non-volatile Memory

Flash memory is the most widely used non-volatile memory technology today. The key device in this prevailing memory is floating gate transistors. A figure of cross section of a floating gate transistor is shown in Figure 1.4. Unlike a MOSFET transistor, an additional floating gate is added between the control gate and channel. Isolated by oxide layers, floating gate is able to trap charges and keep them for years. Therefore, the FG-transistor encodes data based on whether electrons are trapped, and is able to retain data without power. That is where "non-volatile" is derived from.

The principle of read operation can be described as follows. When no charges are trapped in floating gate, the FG-transistor has a threshold voltage of V_{th0}; when negative charges are trapped, they attract positive charges of control gate, thus higher control gate voltage is required to turn on the channel, which produces a higher threshold voltage V_{th1}. By applying intermediate control gate voltage that is between V_{th0} and V_{th1}, and measuring current, the state of device can be known.

The write operation of FG-transistor involves injecting or pulling electrons across the oxide barrier. There are two ways to achieve this: quantum tunneling or hot electron injection. In quantum tunneling scenario, high voltage is applied on control gate, quantum tunneling will take place between the floating gate and channel, and electrons can travel across the oxide barrier. For hot electron injection scenario, electrons are accelerated under high electrical field in the channel till its energy is high enough to penetrate the oxide layer. Note that electrons with high energy

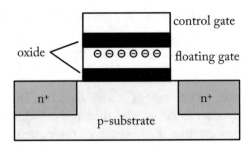

Figure 1.4: The cross section of a floating gate transistor.

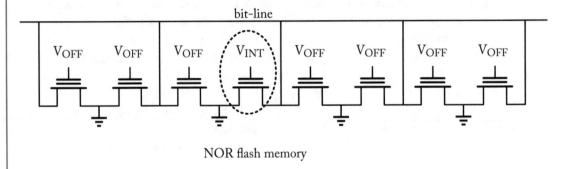

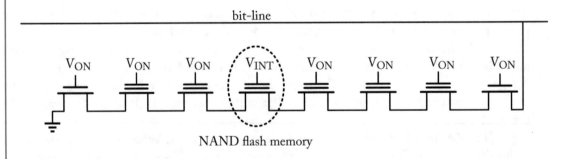

Figure 1.5: Two common layouts for flash memory: NOR flash memory and NAND flash memory.

will damage the oxide lattice, and such damage will accumulate and lead to a limited write cycles, which is typically around 10^5 cycles.

There are two common layouts for flash memory, shown in Figure 1.5: NAND flash memory with FG-transistors in series and NOR flash memory with FG-transistors in parallel. The name NAND and NOR are derived from the fact that their connection fashion in series or parallel resemble a NAND gate or NOR gate. NAND layout has the density advantage over NOR layout because each row only has one ground connection, thus is widely used for external storage.

NOR layout has lower latency, thus it is widely used in embedded systems, where high performance is required. Figure 1.5 also shows how the read of NAND and NOR flash memory can be achieved. The relationship between the different applied voltage magnitude is shown as follows:

$$V_{OFF} < V_{th0} < V_{INT} < V_{th1} < V_{ON} \ll |V_{HIGH}|. \tag{1.2}$$

1.2.2   NANO-SCALE LIMITATIONS

As the technology scaling advances into nano-scale, the conventional memory design faces certain limitations. In the regime where the classic physics still rules, such limitations mainly come from two major aspects. First, due to process variation, the mismatch among transistors may lead to functional failures. Second, positive feedback loop between leakage power and heat may result in thermal runaway failure. In this section, we review the physical mechanisms of such failures induced in nano-scale designs, including write, read, and hold failures, as well as the thermal runaway failure. For the simplification of illustration, only variation of threshold voltage is considered for the functional failures.

Functional Failures by Variation

Write Failure   Write failure is defined as the inability to write data properly into the SRAM cell. During write operation, both access transistors should be strong enough to pull down or pull up the voltage level at internal nodes. As shown in Figure 1.6, write operation can be described on the variable plane as the process of pulling the operating point from initial state (bottom-right corner) to the target state (top-left corner). The crossing line named *separatrix* divides the variable plane into two convergent regions. Given enough time, operating point in any region will converge to the nearest stable equilibrium state either at top-left or bottom-right corner. Write operation is aimed at pulling operating point into a targeted convergent region such that operating point can converge to the closest equilibrium state after operation finishes, as shown by point B in Figure 1.6.

However, an increase in threshold voltage due to variation can reduce the transistor driving strength and vice verse for a decrease in threshold. The increase of V_{th} in M6 along with the decrease of V_{th} in M4 can result in difficulty to pull down v_2. On the variable plane, it becomes more difficult for operating point to move toward the target state. If operating point cannot cross the *separatrix* before access transistors are turned off, it goes back to the initial state, which means a write failure.

Read Failure   Read failure refers to the loss of the previously stored data. Before read operation is performed, both BR and BL are pre-charged to v_{dd}. Suppose previous internal states in SRAM are $v_1 = v_{dd}$ and $v_2 = 0$, electric charge on BR is discharged through M6 and M4 while that on BL remains the same. As such, a small voltage difference between BR and BL is generated which will be detected and amplified. In this way, data stored in the SRAM can be read. Note

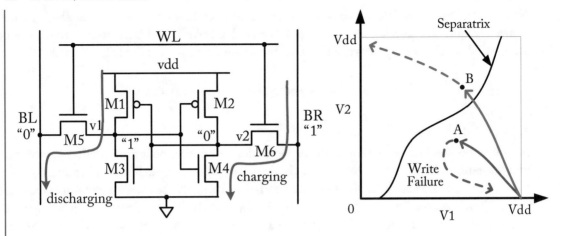

Figure 1.6: Illustration of SRAM write failure.

that access transistors need careful sizing such that their pull-up strength is not strong enough to pull the stored "0" to "1" during read operation.

On the variable plane, operating point is inevitably perturbed and pulled toward the *separatrix*. If read operation does not last too long, access transistors can be shut down before operating point crosses the *separatrix*. As such, the operating point returns to the initial state in the end, as point A in Figure 1.7, which means a read success.

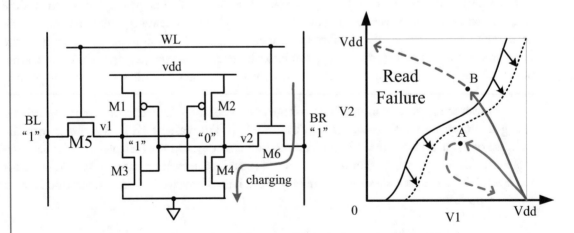

Figure 1.7: Illustration of SRAM read failure.

Even though all the sizing are carefully taken, threshold variations may still result in read failure. For example, variation caused by mismatch between M4 and M6 may result in unbalanced pulling strength, and v_2 can be pulled up more quickly. As a result, operating point crosses the *separatrix* before read operation ends, as point B in Figure 1.7.

Hold Failure   Hold failure happens when the SRAM fails to retain the stored data. It can be caused by external noise or single event upset (SEU). The external perturbation can be modeled as noise current injected into SRAM. Similar to read operation, operating point is expected to converge back to initial state after settling down from disturbance. Otherwise, it will cross to the other convergent region.

While access transistors have no impact on the retention of SRAM data, M1-4 together can determine the likelihood of hold failure by finding the position of the *separatrix* and thus threshold variation may cause failure by perturbing the *separatrix*, as shown in Figure 1.8. As such, one needs to verify if the SRAM is still tolerable to the injected noise in the presence of threshold voltage variation.

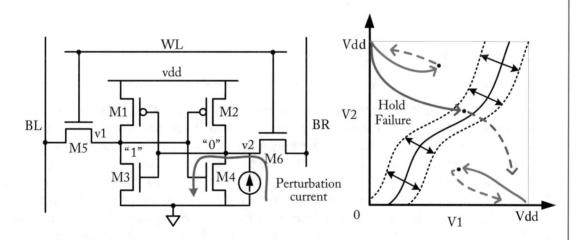

Figure 1.8: Illustration of SRAM hold failure.

Functional Failure by Thermal Runaway

Thermal runaway [Lacey, 1995, Schatz and Bethea, 1994] refers to a situation where an increase in temperature changes the conditions in a way that causes a further increase in temperature in positive feedback way. In a memory system, it is associated with the electrical-thermal coupling between leakage power and temperature. Note that leakage current in memory can be modeled by

$$I_{leakage} = \underbrace{A_s \cdot \frac{W_d}{L_d} \cdot v_T^2(1 - e^{\frac{-V_{DS}}{v_T}}) \cdot e^{\frac{V_{GS}-V_{th}}{ns \cdot v_T}}}_{I_{subthreshold}}$$

$$+ \underbrace{W_d \cdot L_d \cdot A_J(\frac{T_{oxr}}{T_{ox}})^{nt}\frac{V_g \cdot V_{aux}}{T_{ox}^2}e^{-B_J T_{ox}(a-b|V_{ox}|)(1+c|V_{ox}|)}}_{I_{gate\_leakage}},$$

(1.3)

- where $v_T = \frac{k \cdot T}{q}$ is the thermal voltage,

- V_{DD} is the supply voltage with ΔV swing,

- L_d and W_d are the effective device channel length and width,

- ns is the subthreshold swing coefficient,

- A_s, A_J, B_J, a, b, and c are technology-dependent constants,

- nt is a fitting parameter,

- T_{ox} and T_{oxr} are gate dielectric and reference oxide thickness, respectively, and

- V_{aux} is auxiliary temperature-dependent function for density of tunneling carriers.

As the technology node scales down, the controlling ability of transistors becomes weaker, and hence larger leakage current will be experienced. As such, thermal runaway becomes a prominent limitation for large-scale big-data memory integration in advanced technology nodes. The course of potential thermal runaway is illustrated in Figure 1.9. At the very beginning, memory works at room temperature with moderate leakage current, which will consistently produce heat. If the thermal source grows much faster than heat removal ability of heat-sink, there will be thermal accumulation and lead to temperature increase of memory chip. Due to the exponential relationship between temperature and leakage current, the increase of memory temperature will in turn provoke larger leakage current, which in turns increases the leakage current. Such uncontrolled positive feedback will continue and finally lead to the destructive high temperature, melting silicon and permanently damaging the memory cells.

Thermal runaway temperature $T_{threshold}$ is temperature at which thermal runaway failure happens. When the temperature goes beyond $T_{threshold}$, the system temperature will rise rapidly with resulting thermal runaway. As shown in Figure 1.10b, the $T_{threshold}$ represents the maximum heat-removal ability of the heat-sink beyond which thermal runaway failure happens. To avoid thermal runaway, we can either maintain a low thermal source or improve the thermal removal ability. To reduce the thermal source, low power techniques such as power gating or more sophisticated memory cell structures can be applied. For thermal removal, close heat-removal path to heat-sink is favored.

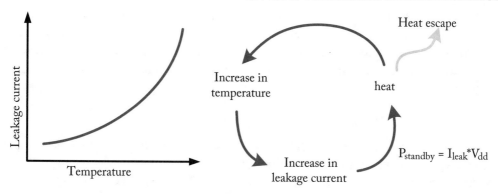

Figure 1.9: Illustration of SRAM thermal runaway failure by positive feedback between temperature and leakage power.

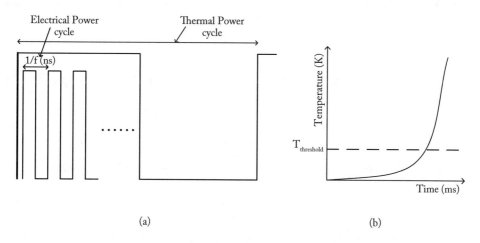

Figure 1.10: (a) Thermal power and electrical power cycles; (b) thermal runaway and threshold temperature.

1.3   NON-VOLATILE SPINTRONIC MEMORY

A large portion of the emerging non-volatile memory devices are magnetization based. The MRAM usually is formed by one insulator in the middle, sandwiched by two ferromagnetic layers, namely fixed layer that is strongly magnetized and free layer that can be easily changed. Differed by the approaches for writing, there are several phases of MRAM technology. The first-generation MRAM needs external magnetic field to switch the free layer magnetization. The second-generation STT-RAM is introduced and the free layer magnetization can be altered by polarized current, which brought significant advantages such as easy integration with current

CMOS technology and high density, high reliability, etc. Recently, the third-generation domain wall racetrack is introduced, with a series of magnetization domains in one ferromagnetic thin-film nanowire, and additional shift ability. The shift is also current-induced operation. In this section, the magnetization dynamics under external field and spin-current in nanosecond regime will be introduced.

1.3.1   BASIC MAGNETIZATION PROCESS

As an intrinsic property, electrons spin about its axis and produce magnetic field like current carrying wire loop. From macrospin point of view, the relation between magnetization M and angular momentum associated with electron spin S can be expressed as

$$M = -\gamma S, \qquad (1.4)$$

where $\gamma = 2.21 \times 10^5$ m A^{-1} s^{-1} is the gyromagnetic ratio. A uniform magnetic field exerts no net force on a current loop but it does exert a net torque, and the torque T, on the current carrying loop under applied magnetic field H, can be expressed as

$$T = M \times H. \qquad (1.5)$$

By definition, the time derivative of angular momentum is called torque. The relation between the angular momentum L, and torque T, reads

$$\frac{dL}{dt} = T. \qquad (1.6)$$

The quantum form of Equation 1.6 still remains valid, and then we have

$$\frac{dS}{dt} = T. \qquad (1.7)$$

By combining Equations 1.4, 1.5, and 1.7, we can obtain the motion equation of magnetization under applied magnetic field

$$\frac{dM}{dt} = -\gamma M \times H. \qquad (1.8)$$

The processional motion described by Equation 1.8 indicates that the magnitude of magnetization will not change, and also the angle between H and M will not change, which is depicted in Figure 1.11. This is based on that no energy loss is assumed during this process.

1.3.2   MAGNETIZATION DAMPING

In real systems, however, energy is dissipated through various means and the magnetization motion is damped until an equilibrium is reached. Energy dissipation can occur through the spin-spin, spin-photon, and spin-electron interactions through which the spin energy is transferred.

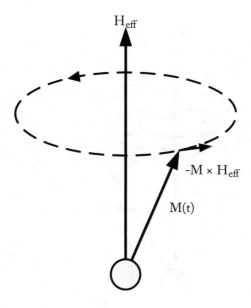

Figure 1.11: The magnetization precession.

The approach followed by Landau and Lifshitz is to introduce dissipation in a phenomenological way. In fact, they introduce an additional torque term that pushes magnetization in the direction of the effective field. Landau-Lifshitz equation in the Gilbert form, or LLG equation, then reads

$$\frac{dM}{dt} = -\gamma M \times H + \frac{\alpha}{M_s} M \times \frac{dM}{dt}. \tag{1.9}$$

The magnetization dynamics described by Equation 1.9 is sketched in Figure 1.12.

1.3.3 SPIN-TRANSFER TORQUE

In 1996, Berger [1996] and Slonczewski [1996] predicted, which later has been confirmed experimentally [Katine et al., 2000, Mojumder et al., 2011, Sun, 1999, Tsoi et al., 1998], that electrons that carry enough angular momentum are able to cause magnetization precession and switching by spin-transfer torque effect. When a current passes through a ferromagnetic layer, its electrons spins are polarized along the magnetization direction and the current becomes spin-polarized and hence carries angular momentum. And the spin-polarized current, when flows through a second ferromagnetic layer, exerts a spin torque on the local magnetic moment of the magnetic layer and causes the magnetization precession and switching when the current is large enough (Figure 1.13).

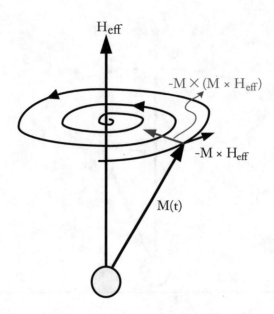

Figure 1.12: The magnetization precession with damping.

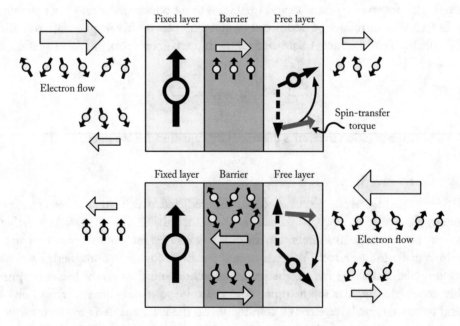

Figure 1.13: The spin-transfer torque effect.

Therefore, the dynamics of the free layer magnetization can be determined by the LLG equation in conjunction with an additional term for spin-transfer torque,

$$\frac{dM}{dt} = -\gamma M \times H + \frac{\alpha}{M_s} M \times \frac{dM}{dt} - \frac{a_J}{M_s}[M \times (M \times P)], \tag{1.10}$$

where P is the magnetization of fixed layer, M_s is the saturation magnetization, and a_J is a factor related to the interfacial interaction between magnetic moment and spin-polarized electrons, which is proportional to the current density, and the sign of a_J depends on the direction of current. When applied properly, the current is able to cancel the damping and switch the magnetization of free layer by spin transfer torque.

For the current induced magnetization precession, there exists a threshold current density J_{c0}, and by applying current larger than J_{c0} the magnetization can be switched back and forth.

$$J_{c0} = \frac{2e\alpha M_s t_F (H_K + H_{ext} + 2\pi M_s)}{\hbar\eta}, \tag{1.11}$$

where e is the electron charge, α the damping constant, M_s is the saturation magnetization, t_F is the thickness of free layer, $\hbar$ is the Planck constant, η is the spin-transfer efficiency, H_K is the anisotropy field, and H_{ext} is the external applied magnetic field.

There are three modes for the current-driven magnetization switching: thermal activation associated with switching time longer than 10 ns, processional switching associated with switching time less than a few nanoseconds, and dynamic reversal as a compound process of both. The above three modes reveal the switching time and current density relationship.

For the fast processional switching, the switching time is reversely proportional to the applied current,

$$\tau_p \propto \frac{1}{(J - J_{c0})} \ln\left(\frac{\pi}{2\theta_0}\right), \tag{1.12}$$

where θ_0 is the initial magnetization angle deviated from the easy axis. At finite temperature, θ_0 is determined by thermal distribution. For the fast processional switching in the regime of nanosecond, it usually takes a current density that is several times greater than J_{c0}.

In the slow thermal activated switch regime, the switching current is dependent on the current pulse width and thermal stability factor $\Delta = K_u V / k_B T$ of the free layer. Interestingly, the current density can be smaller than the critical density, and therefore is useful for current reduction. In this case, the standard thermal agitation will be assisted by spin-current, which introduces extra energy to reach enough energy for the magnetization switching. The relation reads

$$J(\tau) = J_{c0}\left[1 - \frac{K_u V}{k_B T} \ln\left(\frac{\tau}{\tau_0}\right)\right], \tag{1.13}$$

where $\tau_0 \propto 1ns$ is the inverse of the attempt frequency and $K_u V$ is the anisotropy energy.

As the fast processional switching requires large current density which reduces the robustness and cause undesired switching, and the slow thermal activation process takes too much time, the most interesting switching mode is the dynamic reversal at intermediate current pulses. Although the dynamic switching mode corresponds to the operating speed of practical STT-RAM, the explicit formula is hard to be derived due to its complicated process. Therefore, the dynamic reversal is usually studied as a combination of processional and thermally activated switching.

1.3.4   MAGNETIZATION DYNAMICS

Magnetic domains are formed by the competition between the various energy terms involved in a magnetic object. The energy of a magnetic structure is the sum of the exchange energy, the anisotropy energy, Zeeman energy, and the demagnetization energy. The magnetic system seeks to minimize its overall free energy. Since the magnitude of the magnetization cannot change the way to minimize the energy is to vary the direction of the magnetization. The exchange energy seeks to align the spins with each other, the anisotropy energy seeks to align the spins with an axis determined by the crystal structure, the Zeeman energy aligns the spins with an external field. When the magneto static dipole-dipole interaction is also taken into account, known as the demagnetization energy, a nonuniform magnetization will generally be found as the lowest compromise of overall energy. Short-range exchange energy will prevail a configuration with the spins aligned, large-range dipole-dipole interaction will, however, prevail a magnetic state with minimal net magnetization. In the macro-spin model of magnetization dynamics study, the short-range exchange energy can be ignored. The energy associated with anisotropy field can be written as

$$\epsilon = K(1 - m_x^2), \tag{1.14}$$

where K is the anisotropy constant and m_x is the normalized magnetization in x-direction, defined as the in-plane easy axis.

For Zeeman energy by external applied field, we have

$$\epsilon = -\mu_0 M H_{ext} \tag{1.15}$$

in which μ_0 is called the vacuum permeability.

Demagnetization field represents the work necessary to assemble magnetic poles in a given geometrical configuration. Since the thickness is so small compared to the in-plane dimensions, the dominant term is approximately the demagnetizing field of a uniformly magnetized thin film with infinite lateral dimensions, namely $H_D = [0, 0, -M_S m_z]$. The associated energy density is

$$\epsilon = -\frac{1}{2}\mu_0 M H_D = -\frac{1}{2}\mu_0 M_s^2 H_D^2. \tag{1.16}$$

Combining all three terms together, we have the overall energy

$$\epsilon = K(1 - m_x^2) + \frac{1}{2}\mu_0 M_s^2 m_z^2 - \mu_0 M \cdot H_{ext}. \tag{1.17}$$

When pulled out of equilibrium, the magnetization is subject to an effective field. Therefore, we are able to calculate the effective field H in the LLG Equation 1.10,

$$H_{eff} = -\frac{1}{\mu_0 M_s}\frac{\delta\epsilon}{\delta m} = [H_x^{ext} + H_K m_x, 0, -M_s m_z], \tag{1.18}$$

where $H_K = 2K/(\mu_0 M_s)$ is the anisotropy field.

In dimensionless form, we have

$$\omega = \frac{1}{2}Q(1 - m_x^2) + \frac{1}{2}m_z^2 - m \cdot h^{ext} \tag{1.19}$$

and

$$h_{eff} = -\frac{\delta\omega}{\delta m} = [h_x^{ext} + Q m_x, 0, -m_z] \tag{1.20}$$

with $Q = 2K/(m u_0 M_s^2) = H_K/M_s$. Since the magnitude of m does not change, which suggests that it can be rewritten in spherical coordinates,

$$\frac{d\theta}{d\tau} = h_\phi - \alpha\sin\theta\frac{d\phi}{d\tau} \tag{1.21}$$

$$\sin\theta\frac{d\phi}{d\tau} = -h_\theta + \alpha\frac{dehta}{d\tau} \tag{1.22}$$

by multiplying α to Equation 1.21, and add Equation 1.22, and multiplying α to Equation 1.22, and minus Equation 1.22, we could obtain a set of first-order differential equations:

$$(1 + \alpha^2)\frac{d\theta}{d\tau} = h_\phi - \alpha h_\theta \tag{1.23}$$

$$(1 + \alpha^2)\sin\theta\frac{d\phi}{d\tau} = -h_\theta + \alpha h_\phi. \tag{1.24}$$

The stable processional states can be obtained by numerical integration of Equations 1.23 and 1.24, as first demonstrated by Sun [2000]. With approximation, the current threshold for the establishment of a stable magnetization trajectory may be simply derived from standard perturbation theory. Clearly, in the absence of current and under the action of any applied field $h_x^{ext} > 0$, the stable magnetization direction satisfies $m_x = 1$, or, $\theta = \pi/2$, $\phi = 0$. As the damping constant α is generally small, so that in the investigated trajectories zone, the spin-transfer torque roughly balances the damping. Therefore, χ is of the same order of magnitude as α, and both can be treated as small parameters. To investigate the stable processional states, we focus on trajectories in which the magnetization is close to its equilibrium states. This suggests the replacement $\theta = \pi/2 + \xi$, so that ξ and ϕ can be treated as small. Taking $1 + \alpha^2 \cong 1$ leads to the following linearized equations of magnetization motion:

$$\frac{d\xi}{d\tau} = h_\phi - \alpha h_\theta \tag{1.25}$$

$$\frac{d\phi}{d\tau} = -h_\theta + \alpha h_\phi \tag{1.26}$$

with

$$h_\theta = -(1 + Q + h_x^{ext})\xi - \chi\theta \tag{1.27}$$

$$h_\phi = +\chi\xi - (Q + h_x^{ext})\phi. \tag{1.28}$$

Let $u = Q + h_x^{ext}$, and λ be the first derivative operator $d/d\tau$, the first-order differential equation then reads

$$\lambda^2 + [\alpha + 2(\alpha u - \chi)]\lambda + u(1 + u) = 0 \tag{1.29}$$

solution precession

$$\theta = e^{-\frac{t}{t_0}} \cos(\omega t + \Phi_0), \tag{1.30}$$

where

$$t_0 = \frac{1}{\gamma_0 M_s (\chi_{crit} - \chi)} \tag{1.31}$$

$$\omega = \gamma_0 M_s \sqrt{u(1 + u) - (\chi_{crit} - \chi)^2} \tag{1.32}$$

with

$$\chi_{crit} = \alpha(\frac{1}{2} + Q + h_x^{ext}) \cong \frac{\alpha}{2} \tag{1.33}$$

if $Q, h_x^{ext} \ll 1$.

1.3.5   DOMAIN WALL PROPAGATION

Recalling spin-transfer torque effect that when a current is passed through a ferromagnetic material, electrons will polarize, that is, the spin of the conduction electron will align with the spin of the local electrons carrying the magnetic moment of the material. When the conduction electrons subsequently enter a region of opposite magnetization they will eventually become polarized again, thereby transferring their spin momentum to the local magnetic moment, as required by the law of conservation of momentum. Therefore, when many electrons are traversing a domain wall (DW), magnetization from one side of the DW will be transferred to the other side. Effectively the electrons are able to push the DW in the direction of the electron flow i.

The influence of current on DW dynamics is often treated by including two spin torque terms in the LLG equation, see Equation 1.9. When the current, with current density J, is flowing in one direction, the x-direction the LLG equation including the spin torque terms can be written as

$$\frac{\partial M}{\partial t} = -\gamma M \times H + \frac{\alpha}{M_s} M \times \frac{\partial M}{\partial t} - \eta J \frac{\partial M}{\partial x} + \beta\eta J M \times \frac{\partial M}{\partial x}, \tag{1.34}$$

where last two terms are added to the regular LLG equation to describe the effect of current on the magnetization dynamics. The first of these terms expresses the adiabatic spin transfer torque as exerted by a current on magnetic DWs with η the strength of the effect. The second STT term in the equation describes the non-adiabatic current induced effect which relative strength is parameterized by β. The strength of the adiabatic spin torque, η, is widely agreed on Berger [1978], Li and Zhang [2004], Tatara and Kohno [2004], Thiaville et al. [2005], and given by:

$$\eta = \frac{g\mu_B P}{2e M_s},\qquad(1.35)$$

where g is the Land factor, μ_B the Bohr magneton, e the electron charge, M_s the saturation magnetization, and P the electron polarization, all of which the values are very well known except for the electron polarization. Estimates for P range from $P = 0.4$ to $P = 0.7$ [Beach et al., 2008].

1.4 TRADITIONAL MEMORY ARCHITECTURE

Before introducing specific memory technologies, it is important to understand the basic electronic components of which the memory is made up of. A memory chip consists of millions to billions of memory cells, and takes binary address as input and finds target cells correspondingly, so that read and write operations can be performed. In order to efficiently perform this, memory cells are organized in certain fashion, as illustrated in Figure 1.14.

Enormous data cells are divided into multiple data arrays, which are connected by H-tree network. The input address is logically divided into two parts, and will be interpreted respectively. The first part of the address indicates the position of the data array, in which the target cells are kept. The second part of the address reveals the position of the target cells inside the data array. The data array identifier will be used by the predecoders along the H-tree network paths, to route electrical signals to target data array. The most noticeable advantage of H-tree network is that it can ensure an identical signal propagating distance to every data array. This is important to ensure the system is deterministic and the access latency is fixed.

The storage unit in the memory is the data array, whose structure is shown in Figure 1.15. All memory cells lie at the crosspoints of the word-lines and bit-lines. Word-lines and bit-lines are metal wires that propagate signals and will incur certain wire delay due to its parasitic wire resistance and capacitance; therefore, the larger the data array is, the longer access latency can be expected. Each cell stores one bit of information, represented by high or low logic, and its value can be read out by sense amplifiers. If every single cell is directly connect with outside I/O, billions of I/Os will be required which is practically impossible to achieve; therefore, the decoders are used to take binary address to operate on designated cells.

A decoder converts binary address from n input lines to a maximum of 2^n unique output lines. Figure 1.16 shows the circuit of a 2–4 decoder with its truth table. In the memory array, the output lines of word-line decoders are connected to the word-lines, which enable an entire row of data array specified by the address. Because the electrical signals are bidirectional on the bit-line, that is, bit-line can drive cell in write operation and cell can drive bit-line in read operation, the bit-line decoder output lines are connected to the multiplexer, which selectively allows electrical signal of specific column to pass.

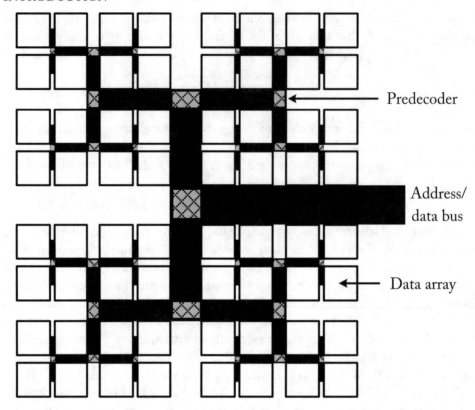

Figure 1.14: Memory organization in H-tree network.

The objective of the readout circuit is to distinguish the bistable states of the memory cells. For conventional memory technologies like static random-access memory (SRAM) and dynamic random-access memory (DRAM), the memory state is represented by electrical voltage: V_{DD} for logic 1 and *GND* for logic 0. Therefore, a readout circuit can be designed as a voltage comparator that compares state voltage with intermediate voltage, namely $V_{DD}/2$. However in practice, due to the charge sharing between memory cell and bit-line parasitic capacitor, the detectable voltage margin is reduced to less than one-tenth of previous value, more sensitive readout circuit is required. Figure 1.17 shows the circuit of a voltage mode typical latch-type sense amplifier, which is able to detect voltage difference smaller than 100 mV at nanosecond scale. Its working mechanism can be described as follows. At first, the *BLP* and *BLN* are precharged at $V_{DD}/2$ and *EN* is kept low.

When the readout is performed, the *BLP* voltage will increase or decrease slightly (ΔV) due to charge share. If differential structure is used, the *BLN* will change $-\Delta V$. Note that the ΔV is determined by the ratio of memory cell capacitance and bit-line capacitance, and when the

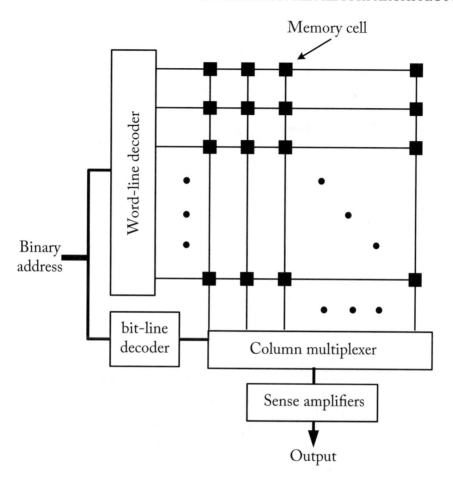

Figure 1.15: The structure of memory array.

array is too large, ΔV will become too small to detect. This is the major reason why the memory cannot be made into one single array and the division is required. After that, the sense amplifier is enabled by set *EN* signal high. This isolates the *BLP* and *BLN* with Q and QB, and Q and QB will be compared by the latch. The latch compares voltages in a positive feedback fashion. Assume there is a ΔV_Q, and as the input of inverter at right hand side ($M3$ and $M4$) it will lower the voltage of V_{QB}, which can be seen in the transfer curve of inverter. And as the input of inverter at left-hand side ($M1$ and $M2$), the decrease in V_{QB} will in turn increase the value of ΔV_Q. As such, the two cross-coupled inverters reinforce each other and enter a positive feedback loop until they reach the final stable state that the ΔV_Q is V_{DD} and V_{QB} is 0. Without the pass transistor $M5$ and

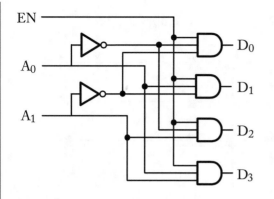

EN	A_0	A_1	D_0	D_1	D_2	D_3
1	0	0	1	0	0	0
1	0	1	0	1	0	0
1	1	0	0	0	1	0
1	1	1	0	0	0	1
0	X	X	0	0	0	0

Figure 1.16: 2–4 decoder logic and truth table.

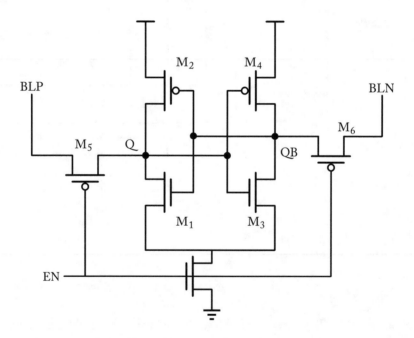

Figure 1.17: Latch-type sense amplifier.

$M6$, the latch will have to drive the entire bit-line, which greatly affects the convergence speed and incur more energy consumption.

1.5   NON-VOLATILE IN-MEMORY COMPUTING ARCHITECTURE

Conventionally, all the data is maintained within memory that is separated from the processor but connected with I/Os. Therefore, during the execution, all data needs to be migrated to processor and written back afterward. In the data-oriented applications, however, this will incur significant I/O congestions and hence greatly degrade the overall performance [Burger et al., 1996]. In addition, significant standby power will be consumed in order to hold the large volume of data.

Theoretically, it is feasible to overcome the bandwidth issue by adding more I/O pins or operating them at higher frequency. Practically, however, the I/O frequency is limited by the signal propagation delay and signal integrity issues, and I/O number is limited by the packaging technology, thus the bandwidth can hardly get further improved.

Instead of improving memory bandwidth, it is also possible to reduce the required data communication traffic between memory and processor. The basic idea behind is that, instead of feeding processor large volume of raw data, it is beneficial to preprocess the data and provide processor only intermediate result. The key to lower communication traffic is the operands reduction. For example, to perform a sum of ten numbers, instead of transmitting ten numbers to processor, in-memory architecture is able to obtain the sum by in-memory logic and transmit only one result thus reduce traffic by 90%. To perform in-memory logic, it is necessary to implement logic inside memory so that preprocessing logic can be done. Such architecture is called logic-in-memory architecture. In general, the requirement of the ideal logic-in-memory architecture can be summarized in Figure 1.18. A big non-volatile memory sea is connect with thousands of small accelerator cores through high bandwidth and energy-efficient reconfigurable I/Os. Considering the leakage reduction at the same time, logic-in-memory architectures that are associated with non-volatile memory are presented in Kautz [1969], Kimura et al. [2004], Matsunaga et al. [2008, 2009], and Hanyu et al. [1998]. Figure 1.19 shows logic in memory architecture at memory cell level. The example illustrated here is an in-memory full adder with both *sum* logic and *carry* logic.

The basic circuitry, including access transistor, the word-line and bit-lines, is to ensure memory access. The data is stored in non-volatile memory devices which have either low or high resistance. Redundant data is required for each bit of data for logic purpose. Combinational logic circuit is added inside which the non-volatile devices are equivalent to transistors: considered turned on if at low-resistance state or turned off if at high-resistance state. In such architecture, the desired result can be obtained immediately without reading operands as if the results are already stored in data array and it is just be "readout". This is very useful for some specific applications as this architecture is able to preprocess data without loading data to processor with extremely short latency.

As the logic is inserted to one cell or a few cells, it is limited to small size thus cannot be made complex. Usually, only simple logic is suitable for such architecture otherwise the overhead would be overwhelming. Although simple logic in such architecture is able to share the workload of processor, its effect to reduce communication traffic is not obvious due to limited operands

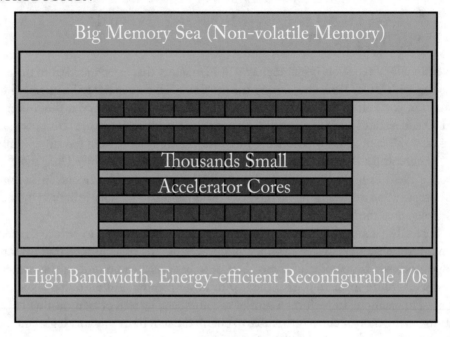

Figure 1.18: Ideal logic-in-memory architecture.

reduction. In addition, similar to the operation of memory, for the whole data array only a few logic can be active concurrently at one time. This leads many logic circuits to be idle at most of the time, which is not only a waste of computational resources but also incurs leakage power for CMOS logic. Another disadvantage is that the data needs to be stored in a very strict manner, determined by in-memory logic circuit functionality.

An alternative in-memory architecture we are presenting at block level in distributed fashion is illustrated in Figure 1.20, which is more effective for traffic reduction. A memory data is usually organized in H-tree fashion, and the data block can be the data array or a number of data arrays that belong to same "H-tree" branch. Instead of inserting in-memory logic at memory cell level inside the data array, the architecture in Figure 1.20 pairs each block of data with in-memory logic (accelerators). Different from the cell level in-memory architecture, the accelerators can be made with higher complexity, and the number of accelerators for each data block can also be customized. The data flow of the block level in-memory architecture is to readout data from data block to in-memory logic, which performs particular functionality and then writes back the result. The data also needs to be stored in assigned blocks but it is much more flexible than that of cell level in-memory architecture. The block level in-memory architecture is very effective to reduce communication traffic between memory and processor. This is because significant operands reduction can be achieved by deploying accelerator with high level functionality. For example, for face recognition in image processing application, instead of transmitting a whole image to ob-

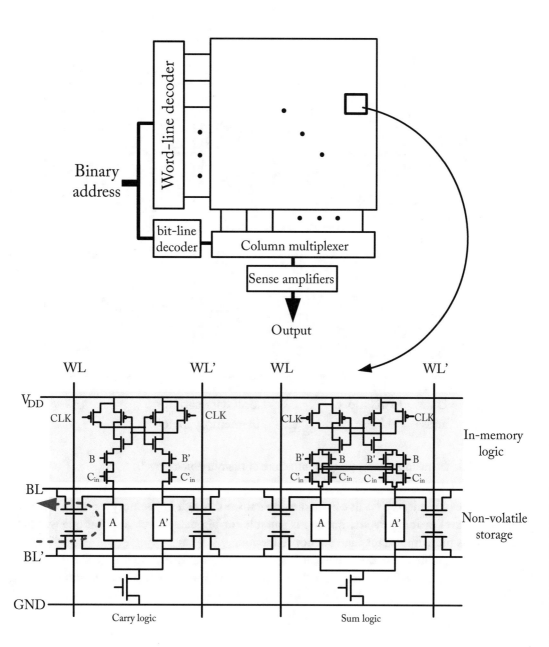

Figure 1.19: In-memory computing architecture at memory cell level.

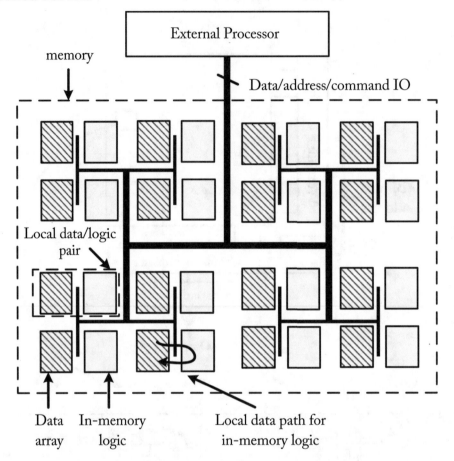

Figure 1.20: In-memory computing architecture at memory block level.

tain a Boolean result, the result can be directly gained through in-memory logic. In other words, the block level in-memory architecture is suitable for big data driven applications where traffic reduction is more important than latency reduction.

In this architecture, the non-volatile spintronic is intensively used. Both the memory block and logic block in each pair are purely implemented by spintronic devices. In addition, energy-efficient in-memory logic units are deployed in the external processor to execute instructions that cannot be accelerated by in-memory logic.

1.6   REFERENCES

Beach, G., Tsoi, M., and Erskine, J. (2008) Current-induced domain wall motion. *Journal of Magnetism and Magnetic Materials*, 320(7), pp. 1272–1281. DOI: 10.1016/j.jmmm.2007.12.021.

21

Berger, L. (1978) Low-field magnetoresistance and domain drag in ferromagnets. *Journal of Applied Physics*, 49(3), pp. 2156–2161. DOI: 10.1063/1.324716. 21

Berger, L. (1996) Emission of spin waves by a magnetic multilayer traversed by a current. *Physical Review B*, 54(13), p. 9353. DOI: 10.1103/physrevb.54.9353. 15

Burger, D., Goodman, J. R., and Kägi, A. (1996) Memory bandwidth limitations of future microprocessors, vol 24, *ACM*. DOI: 10.1145/232974.232983. 25

Hanyu, T., Teranishi, K., and Kameyama, M. (1998) Multiple-valued logic-in-memory vlsi based on a floating-gate-mos pass-transistor network. In *Solid-State Circuits Conference, Digest of Technical Papers*. IEEE International, pp. 194–195. DOI: 10.1109/isscc.1998.672432. 25

Katine, J., Albert, F., Buhrman, R., Myers, E., and Ralph, D. (2000) Current-driven magnetization reversal and spin-wave excitations in co/cu/co pillars. *Physical Review Letters*, 84(14), p. 3149. DOI: 10.1103/physrevlett.84.3149. 15

Kautz, W. H. (1969) Cellular logic-in-memory arrays. *Computers, IEEE Transactions on*, 100(8), pp. 719–727. DOI: 10.1109/t-c.1969.222754. 25

Kimura, H., Hanyu, T., Kameyama, M., Fujimori, Y., Nakamura, T., and Takasu, H. (2004) Complementary ferroelectric-capacitor logic for low-power logic-in-memory vlsi. *Solid-State Circuits, IEEE Journal of*, 39(6), pp. 919–926. DOI: 10.1109/jssc.2004.827802. 25

Lacey, A. (1995) Thermal runaway in a non-local problem modeling ohmic beating: Part 1: Model derivation and some special cases. *European Journal of Applied Mathematics*, 6(2), pp. 127–144. DOI: 10.1017/s095679250000173x. 11

Li, Z. and Zhang, S. (2004) Domain wall dynamics and spin-wave excitations with spin-transfer torques. *Physical Review Letters*, 92(20), pp. 207–203. DOI: 10.1103/physrevlett.92.207203. 21

Matsunaga, S., Hayakawa, J., Ikeda, S., Miura, K., Hasegawa, H., Endoh, T., Ohno, H., and Hanyu, T. (2008) Fabrication of a nonvolatile full adder based on logic-in-memory architecture using magnetic tunnel junctions. *Applied Physics Express*, 1(9), p. 1301. DOI: 10.1143/apex.1.091301. 25

Matsunaga, S., Hayakawa, J., Ikeda, S., Miura, K., Endoh, T., Ohno, H., and Hanyu, T. (2009) Mtj-based nonvolatile logic-in-memory circuit, future prospects and issues. In *Proc. of the Conference on Design, Automation and Test in Europe*, European Design and Automation Association, pp. 433–435. DOI: 10.1109/date.2009.5090704. 25

Mojumder, N. N., Gupta, S. K., Choday, S. H., Nikonov, D. E., and Roy, K. (2011) A three-terminal dual-pillar stt-mram for high-performance robust memory applications. *Electron Devices, IEEE Transactions on*, 58(5), pp. 1508–1516. DOI: 10.1109/ted.2011.2116024. 15

Schatz, R. and Bethea, C. (1994) Steady state model for facet heating leading to thermal runaway in semiconductor lasers. *Journal of Applied Physics*, 76(4), pp. 2509–2521. DOI: 10.1063/1.358509. 11

Slonczewski, J. C. (1996) Current-driven excitation of magnetic multilayers. *Journal of Magnetism and Magnetic Materials*, 159(1), pp. L1–L7. DOI: 10.1016/0304-8853(96)00062-5. 15

Sun, J. (1999) Current-driven magnetic switching in manganite trilayer junctions. *Journal of Magnetism and Magnetic Materials*, 202(1), pp. 157–162. DOI: 10.1016/s0304-8853(99)00289-9. 15

Sun, J. (2000) Spin-current interaction with a monodomain magnetic body: A model study. *Physical Review B*, 62(1), p. 570. DOI: 10.1103/physrevb.62.570. 19

Tatara, G. and Kohno, H. (2004) Theory of current-driven domain wall motion: spin transfer vs. momentum transfer. *Physical Review Letters*, 92(8), pp. 086,601. DOI: 10.1103/physrevlett.92.086601. 21

Thiaville, A., Nakatani, Y., Miltat, J., and Suzuki, Y. (2005) Micromagnetic understanding of current-driven domain wall motion in patterned nanowires. *EPL (Europhysics Letters)*, 69(6), p. 990. DOI: 10.1209/epl/i2004-10452-6. 21

Tsoi, M., Jansen, A., Bass, J., Chiang, W. C., Seck, M., Tsoi, V., and Wyder, P. (1998) Excitation of a magnetic multilayer by an electric current. *Physical Review Letters*, 80(19), p. 4281. DOI: 10.1103/physrevlett.80.4281. 15

CHAPTER 2

Non-volatile Spintronic Device and Circuit

Abstract   The bistable states are the foundation of all memory devices to store data. For conventional memory devices, the bistable states are represented by voltage levels and the transition is described by the charging and discharging of the capacitors. The transition dynamics is critical in order to obtain important figures of merit such as device operation speed and energy. Therefore, it is of great importance to quantitatively understand the physical mechanism and transition dynamics of the emerging non-volatile devices, whose states are represented by non-electrical variables. For the magnetoresistive random-access memory family, including STT-MRAM, and racetrack memory, the magnetization dynamics is the fundamental physics behind, while for the resistive random-access memory category, including memristor and CBRAM, the ion migration effect is the shared physics. In this chapter, both the magnetization dynamics and ion migration dynamics are introduced. In addition, spintronics-based storage and logic implementation is introduced.

2.1   SPICE FORMULATION WITH NEW NANO-SCALE NVM DEVICES

In order to deal with a design composed of large number of non-volatile memory (NVM) devices but also the other traditional devices such as CMOS transistors, the new NVM element needs to be included into a circuit simulator like SPICE [Nagel and Pederson, 1973] with state explicitly described. Traditional nodal analysis (NA) only contains nodal voltages at terminals of devices. Since an inductor is short at *dc* and its two terminal voltages become dependent, the state matrix is indefinite at *dc*. This problem is resolved by a modified nodal analysis (MNA) [Ho et al., 1975], which modifies the NA by adding branch currents as state variables. However, many non-traditional devices (memristor, spin-transfer torque device, etc.) introduced at the nano-scale have to be described by state variables different from the traditional nodal voltages and branch currents, i.e., electrical states. As such, the conventional circuit formulation in MNA may not be able to include these new nano-devices, which contain the non-electrical or non-volatile states [Fei et al., 2012, Shang et al., 2012].

Due to the lack of development of related circuit simulator, the behavior of NVM devices is currently studied in very limited scalability. The challenges faced when integrating with CMOS devices remain unsolved. With the aid of one SPICE-like simulator for non-volatile memory devices developed in this chapter to describe both electrical and non-electrical states, the hybrid CMOS/NVM co-simulation can be efficiently conducted with high accuracy.

2.1.1   TRADITIONAL MODIFIED NODAL ANALYSIS

Kirchhoff's Current Law (KCL) and Kirchhoff's Voltage Law (KVL) are two fundamental equations governing the electric property of a circuit. These two laws can be compactly formulated by an incidence matrix determined by the topology of circuits. Assuming n nodes and b branches, the incident matrix $E \in (R^{n \times b})$ is defined by

$$e_{i,j} = \begin{cases} 1, & \text{if branch } j \text{ flows into node } i \\ -1, & \text{if branch } j \text{ flows out of node } i \\ 0, & \text{if branch } j \text{ is not included at node } i. \end{cases} \quad (2.1)$$

By further denoting branch current as j_b, branch voltages as v_b and nodal voltages as v_n, KCL and KVL can be described by (Table 2.1)

$$\begin{cases} E j_b = 0, & KCL \\ E^T v_n = v_b, & KVL. \end{cases} \quad (2.2)$$

Table 2.1: Definitions of variables used for spice-like simulator

Variables	Definitions
S_m	Non-volatile state variable for new MNA
v_n, j_b	Traditional state variables for MNA: nodal voltage and branch current
v_b, j_i, j_l	Branch voltage, source current, and inductor current
G, C, L_l	Traditional conductance, capacitance, and inductance
S	New Jacobian as memductance
$K_v^F K_s^F K_v^G K_s^G$	New Jacobians introduced by s_m
E_c, E_g, E_l, E_i	Incident matrix for capacitor, resistor, inductor, and current source
E_m	New incident matrix introduced for NVM devices
f, g	Functions introduced by s_m for new state equations

Ideally, the branch current vector is a function purely dependent on the nodal voltages under the device branch equation:

$$j_b = \frac{d}{dt} q \left(E^T v_n, t \right) + j(E^T v_n, t). \quad (2.3)$$

However, as inductor and voltage source become indefinite at dc when using the nodal voltages only (NA), the MNA breaks the branch current vector into four pieces with four corresponding incident matrices, and deploys branch inductive current j_l and branch source current

j_i as new state-variables. As such, the KCL and KVL in Equation 2.2 become

$$\begin{cases} \dfrac{d}{dt} E_c q \left(E_c^T v_n, t \right) + E_g j \left(E_g^T v_n, t \right) + E_l j_l + E_i j_i = 0, \\ L_l \dfrac{d}{dt} j_l - E_l v_n = 0, \\ E_i^T v_n = 0. \end{cases} \tag{2.4}$$

Here the four incident matrices $[E_c, E_g, E_l, E_i]$ describe the topological connections of capacitive, conductive, inductive, and voltage-source elements. Introducing the state variable $x = [v_n, j_l, j_i]^T$, the above MNA formulation can be denoted shortly by a differential-algebra-equation (DAE) below,

$$F(x, \dot{x}, t) = \frac{d}{dt} q(x, t) + j(x, t) = 0. \tag{2.5}$$

2.1.2 NEW MNA WITH NON-VOLATILE STATE VARIABLES

In order to handle the dynamic models of NVM device with non-volatile state variables, one needs to develop a new MNA by adding non-volatile state variables into the traditional MNA (for CMOS devices). As shown in Figure 2.1b, these non-volatile state variables, termed as s_m, determine the conductance of all non-volatile memory devices, termed as memductance (memory conductance) here, and therefore can be categorized as one new device branch. Note that the simulation time is directly related to the total number of state variables generated for the circuit. Compared to the traditional approaches with complex equivalent circuits, the introduction of memductance adds much fewer state variables by including non-volatile state variables to represent dynamic effect of NVM device. Take memristor for example; its equivalent circuit requires dozens of additional nodal voltages to characterize the circuit behavior, while only two nodal voltages and one non-electrical state variable are required in our memductance approach. Therefore, the introduction of memductance greatly simplifies the model complexity and in turn largely reduces the verification and design cost.

One NVM device may require one or multiple non-volatile state variables to accurately describe its dynamic behaviors. The corresponding incident matrix is termed as E_m, with which the nonvolatile device branch current is obtained as $E_m j(E_m^T v_n, s_m, t)$. As Figure 2.1b shows, combined with branch currents from the traditional CMOS devices, a new *KCL* equation is formed by

$$\begin{cases} \dfrac{d}{dt} E_c q \left(E_c^T v_n, t \right) + E_m j \left(E_m^T v_n, s_m, t \right) + E_g j \left(E_g^T v_n, t \right) + E_l j_l + E_i j_i = 0, \\ L_l \dfrac{d}{dt} j_l - E_l v_n = 0, \\ E_i^T v_n = 0, \\ f(E_m^T v_n, s_m, t) + \dfrac{d}{dt} g(E_m^T v_n, s_m, t) = 0. \end{cases} \tag{2.6}$$

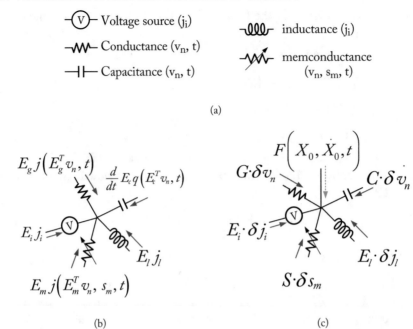

Figure 2.1: New MNA with (a) components and state variables; (b) large signal *KCL*; and (c) small signal *KCL*.

Here functions f and g are the additional state equations for nonvolatile devices with s_m. Moreover, with the new state variable vector $X = [v_n, j_i, j_l, s_m]^T$, the above new MNA can still be described by the differential-algebra-equation (DAE) as Equation 2.5.

We can further derive the Jacobian as generalized conductance G, capacitance C, and memductance S. The additional term of memductance S is introduced to describe the conductance of nonvolatile devices for the induced current change under the change of non-volatile state s_m. At one biasing point $X0$, the first-order derivative or Jacobian of the new DAE with respective to X are

$$
\begin{cases}
G = \left(E_g \dfrac{d}{dv_b^g} j\left(v_b^g, t\right) E_g^T + E_m \dfrac{d}{dv_b^m} j\left(v_b^m, s_m, t\right) E_m^T \right) ||_{X=X_0}, \\[2mm]
C = \left(E_c \dfrac{d}{dv_b^c} q\left(v_b^c, t\right) E_c^T \right) ||_{X=X_0}, \\[2mm]
S = \left(E_m \dfrac{d}{ds_m} j\left(v_n^m, s_m, t\right) E_m^T \right) ||_{X=X_0},
\end{cases}
\tag{2.7}
$$

where $v_b^g = E_g^T v_n$, $v_b^m = E_m^T v_n$, $v_b^c = E_c^T v_n$.

In addition, there are four additional Jacobian terms introduced from functions f and g due to the new state variable s_m for NVM device:

$$
\begin{cases}
G = \left(E_g \dfrac{d}{dv_b^g} j\left(v_b^g, t\right) E_g^T \right) ||_{X=X_0}, v_b^g = E_g^T v_n, \\[2mm]
C = \left(E_c \dfrac{d}{dv_b^c} q\left(v_b^c, t\right) E_c^T \right) ||_{X=X_0}, v_b^c = E_c^T v_n, \\[2mm]
S = \left(E_l \dfrac{d}{d\Phi_b^l} j\left(\Phi_b^l, t\right) E_l^T \right) ||_{X=X_0}, \Phi_b^l = E_l^T \Phi_n, \\[2mm]
W = \left(E_m \dfrac{d}{d\Phi_b^m} q\left(\Phi_b^m, t\right) E_m^T \right) ||_{X=X_0}, \Phi_b^m = E_m^T \Phi_n.
\end{cases}
\tag{2.8}
$$

The subscripts v and s denote the derivatives to nonvolatile device branch voltage v_b^m and its non-volatile state variable s_m, respectively. The superscript F and G correspond to functions f and g, which are current and charge terms for the KCL equation, respectively.

In addition, with the new small-signal current introduced by s_m as shown in Figure 2.1c, the new linearized small-signal DAE becomes

$$
\begin{cases}
G \cdot \delta v_n + C \cdot \delta \dot{v}_n + E_i \cdot \delta j_i + E_l \cdot \delta j_l + S \cdot \delta s_m = -F\left(X_0, \dot{X}_0, t\right), \\[1mm]
E_i^T \cdot \delta v_n = 0, \\[1mm]
E_l^t \cdot \delta v_n = L_l \cdot \delta \dot{j}_l, \\[1mm]
K_v^F \cdot \delta v_n + K_s^F \cdot \delta s_m + K_v^G \cdot \delta \dot{v}_n + K_s^G \cdot \delta \dot{s}_m = 0.
\end{cases}
\tag{2.9}
$$

Therefore, we have the following matrix formed for linearized system equation with all Jacobian matrices ($G, C, L_l, S, K_v^F, K_s^F, K_v^G, K_s^G$) and incident matrices (E_c, E_g, E_l, E_i):

$$
\begin{bmatrix}
G & E_i & E_l & S \\
-E_i & 0 & 0 & 0 \\
-E_l & 0 & 0 & 0 \\
K_v^F & 0 & 0 & K_s^F
\end{bmatrix}
\begin{bmatrix}
\delta v_n \\
\delta j_i \\
\delta j_l \\
\delta s_m
\end{bmatrix}
+
\begin{bmatrix}
C & 0 & 0 & 0 \\
0 & 0 & 0 & 0 \\
0 & 0 & L_i & 0 \\
K_v^G & 0 & 0 & K_s^G
\end{bmatrix}
\begin{bmatrix}
\delta \dot{v}_n \\
\delta \dot{j}_i \\
\delta \dot{j}_l \\
\delta \dot{s}_m
\end{bmatrix}
= -F(X_0, \dot{X}_0, t).
\tag{2.10}
$$

Note that both large-signal and small-signal system equations can be obtained when introducing non-volatile state variables. Its implementation for NVM devices is shown in the following section.

2.2 STT-MTJ DEVICE AND MODEL

2.2.1 STT-MTJ

Based on spin-transfer torque effect, STT-RAM is seen as the second phase of magnetization based on non-volatile memory technology after the toggle MRAM. Since it has great scalability thanks to current-induced magnetization switch instead of external magnetic field, it has attracted a lot of attention. In the following, the non-volatile state variables of STT-MTJ are identified based on the magnetization physics.

Non-volatile State Identification

In this section, the operation of STT-MTJ device is discussed. Then the non-volatile state variables of STT-MTJ devices are identified for magnetization angles with the consideration of arbitrary driving condition. All variables used in this section are summarized in Table 2.2.

Table 2.2: Definitions of variables used for STT-MTJ device

Variables	Definitions
θ, ϕ	Shown in Figure 2.2, azimuthal angles of magnetization orientation in x-z and x-y plane
R_H, R_L	Resistance values of anti-parallel state and parallel state
ΔR_{GMR}	Difference between R_H and R_L
α	Damping constant
e	Electron charge
A	Area of STT-MTJ cross-section
l_m	Thickness of oxide barrier
H^{ext}, H_k	External applied field and shape anisotropy field
M	Magnetization
M_s	Saturation magnetization of material
m, h	Normalized magnetization and effective magnetic field
η	Spin transfer efficiency
$\hbar$	Reduced Planck constant
τ	Normalized time, $\tau = \gamma_0 M_s t$
τ_0	Inverse of attempt frequency (1ns)
Δ	Thermal stability factor
I_c	Critical current for magnetization switching

A typical STT-MTJ device structure appears as a sandwich with two ferromagnetic layers and one oxide barrier in between [Brinkman et al., 1970, Wang et al., 2009]. One STT-MTJ device has two stable states: parallel state (P) or anti-parallel state (AP), where the free layer magnetization is in the same or the opposite direction with hard axis (magnetization direction of fixed layer), respectively. One giant magneto resistance (*GMR*) at AP state (R_H) is higher than the *GMR* at P state (R_L) [Grünberg et al., 1986]. The angle θ is one magnetization angle between

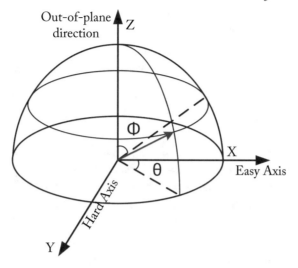

Figure 2.2: Spherical coordinates with two magnetization angles: θ and ϕ.

free layer and hard axis, and the *GMR* can be expressed as [Baibich et al., 1988]:

$$\begin{cases} R(\theta) = R_L + \dfrac{R_H - R_L}{2}(1 - \cos(\theta)) \\ = R_L + \dfrac{\Delta R_{GMR}}{2}(1 - \cos(\theta)). \end{cases} \tag{2.11}$$

It can be easily derived from Equation 2.11 that: $R(\theta = 0) = R_L$ at P state and $R_{(\theta} = \pi) = R_H$ at AP state.

The operating principle of STT-MTJ device can be summarized as: the external current induces the state (P or AP) change. STT-MTJ is switched from P to AP with sufficient forward biased current and from AP to P with sufficient reverse biased current. In order to model this mechanism accurately, we need to look into two dominant effects in the device physics of STT-MTJ: tunneling effect [Brinkman et al., 1970] and spin transfer torque effect.

Tunneling effect of STT-MTJ can be understood as a parabolic relationship between the junction conductance and the applied voltage [Brinkman et al., 1970]. Tunneling effect normally becomes dominant when the applied voltage is relatively small such that it will not trigger the STT-MTJ to change state. This relation can be approximated by:

$$\begin{cases} R_l(V) = \dfrac{R_{l0}}{1 + c_l V^2} \\ R_h(V) = \dfrac{R_{h0}}{1 + c_h V^2}, \end{cases} \tag{2.12}$$

where c_l and c_h are voltage-dependent coefficients for parallel state and anti-parallel states, respectively.

As introduced in Chapter 2, spin transfer torque effect is able to cause magnetization reversal in free layer of STT-MTJ when the spin-polarized current is larger than the critical value I_c. Magnetization reversal is not an instantaneous process, and the the switching time required (T_s) for magnetization reversal decreases exponentially with the current applied (I_0),

$$T_s = \tau_0 \exp\left(\Delta\left(1 - \frac{I_0}{I_c}\right)\right), \tag{2.13}$$

where $I_c = (2eA\alpha l_m M_s(H + H_k + 2\pi M_s))/\eta$, τ_0 is the switching time at $I_0 = I_c$, and the definitions of remaining variables are in Table 2.2. Note that T_s from Equation 2.13 is the minimum pulse width requirement for certain I_0. For the write operation in STT-MTJ device-based memory, one must make sure that the constant current larger than I_c is applied within a period of T_s.

To better understand T_s is important for modeling STT-MTJ devices. This requires the analysis for the dynamic behavior of STT-MTJ device under arbitrary driving condition. As described in Chapter 2, the Landau-Lifshiltz-Gilbert equation (LLG) is deployed for this purpose, and the dynamics can be expressed as:

$$\theta = \theta_0 Exp\left(-\frac{t}{t_0}\right) \cdot \cos(\phi) \tag{2.14}$$

$$\omega = \frac{d\phi}{dt} = k_1 \sqrt{k_2 - (k_3 - k_4 I)^2}, \tag{2.15}$$

where θ_0 is the initial value of θ, slightly tilted from the stable x or $-x$ directions; t_0 is procession time constant; ω is the angular speed of ϕ; k_1 to k_4 are magnetic parameters with detailed explanation in Chapter 2; and I is the spin-current that causes the magnetization precession.

Simulation Results

The 1T-1MTJ structure STT-RAM array with transient analysis for a write operation is set as the test bench circuit. The circuit netlist is described in two versions which only differ in terms of STT-MTJ model, one using the equivalent circuit- (Figure 2.3) based SPICE macromodel in Harms et al. [2010], and one using the intrinsic physics-based model in NVM-SPICE. Necessary modifications are made to port the HSPICE subcircuit netlist provided by Harms et al. [2010] compatible with Berkeley SPICE styled NVM-SPICE. The simulation duration is set 20 ns with time step of 0.1 ns.

Table 2.3 shows the runtime comparison of the two different simulation approaches for different array sizes. It can be observed that the simulation using NVM-SPICE intrinsic physics-based model is 10x~117x faster than the equivalent circuit approach. The advantage will be even larger when the array size increases. For typical memory array size of hundreds by hundreds, the speedup can be expected more than 100x.

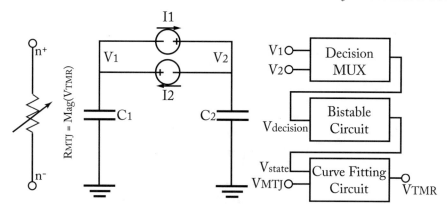

Figure 2.3: Equivalent circuit of STT-MTJ model.

Table 2.3: Simulation time comparison for STT-RAM array using different simulation approaches (unit in second)

Array Size	Macromodel [Harms et al., 2010]	NVM-SPICE	Speedup
8 × 8	2.522	0.257	10x
16 × 16	98.131	1.87	52x
32 × 32	1119.99	11.533	97x
64 × 64	22188.8	189	117x

2.2.2   STT-RAM

STT-RAM [Hosomi et al, 2005, Kawahara et al, 2007] is based on the spin-transfer torque effect, in which magnetization orientations (corresponding to low-high resistance states) in magnetic multi-layer nano-structures can be manipulated via spin polarized current. The spin transfer phenomena occur for the electric current flowing through two ferromagnetic layers separated by a thin nonmagnetic spacer layer. The current becomes spin polarized by transmission through or reflection from the first ferromagnetic layer (the pinned reference layer) and mostly maintains this polarization as it passes through the nonmagnetic spacer and enters and interacts with the second ferromagnetic layer (the free layer). This interaction exerts a spin torque on the magnetic moment of the free layer (FL) through a transfer of angular momentum from the polarized current to the FL magnetization. This spin torque can oppose the intrinsic damping of the FL causing the magnetization precession (exiting spin waves) or reverse the direction of the magnetization with sufficient current strengths. Spin-transfer can have important implications on electronic device applications since it provides a local means of magnetization manipulation rather than using the long-range Oersted field generated by a remote current.

This new MRAM design, called spin-transfer torque RAM (STT-RAM), is believed to have better scalability than conventional MRAM because its switching current is proportional to the MTJ size (Figure 2.4).

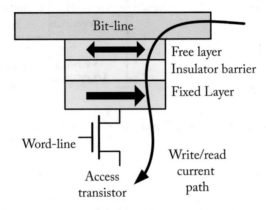

Figure 2.4: The diagram of spin-transfer torque-based MRAM cell structure.

2.2.3   TOPOLOGICAL INSULATOR

Topological insulator (TI) is recently discovered nano-device whose bulk acts as insulator but surface behaves as metal. As state information in a TI device is conducted by ordered spins, it draws tremendous interest for ultra-low power computing.

The scattering, in which electrons deviate from their trajectory resulting in dissipation, is the fundamental reason of power consumption. As such, instead of manipulating electrons, it is envisioned that spintronics [Wang et al., 2008, Wolf et al., 2001] can be developed by controlling pure spin current and spin accumulation for the ultra-low power information storage, transmission and processing. Topological insulator (TI) [Bernevig et al., 2006, Fu et al., 2007, Hsieh et al., 2008, Moore, 2010] is recently found as a means for realizing spintronics not only for the underpinning of computer logic in microprocessors but also for hard-disk, written, read, or rewritten with significantly reduced power.

Materials like Bi_2Se_3, Bi_2Te_3, and Sb_2Te_3 are experimentally observed as three-dimensional (3D) TI devices [Chen et al., 2009, Hsieh et al., 2009, Xia et al., 2009]. TI has insulating band-gap state in the bulk and gap-less metallic state at surface. The gap-less metallic state at surface in a 3D TI is very robust under perturbations. Due to the strong spin-orbit coupling, electrons in TI move along their surface into two distinguished directions without scattering according to their spins. This works similarly to that vehicles can move in two opposite directions at two sides of a highway without disturbing each other. Without scattering, power and further thermal dissipation can be significantly reduced in this type of devices.

The quantum spin Hall effect (QSHE) behavior has been observed for TI devices [Brüne et al., 2011, Cheng et al., 2010, Zyuzin and Burkov, 2011], where a quantized topological surface states form Landau levels in presence of external magnetic field. This becomes the foundation for its application in magnetic non-volatile memory (NVM) design [Fujita et al., 2011]. However, no design exploration is performed before on the potential use of TI in NVM. From a modeling perspective, due to its unique device properties, there is no physical model to fully capture the states in TI devices. From design perspective, there is no NVM design based on TI on how to apply the external field for read and write operation. There is a need to develop a design platform for TI based NVM design, including SPICE-like simulation of TI device and its according memory design.

Non-volatile State Identification

A spin-transfer torque magnetic tunneling junction (STT-MTJ) device is found with state depending on magnetization angle. Similarly, a topological insulator (TI) device also has non-conventional electrical states to describe. In this section, the working mechanism of TI is discussed with additional state variable identified as well. Then, similar to the BSIM model for MOSFET, the according device model to stamp a TI device in SPICE is described. The deployed variables and terms are summarized in Table 2.4.

Table 2.4: Definitions of variables used for TI device

Variables	Definitions
θ, ϕ	Shown in Figure 2.2, azimuthal angles of magnetization orientation in x-z and x-y plane
α	Damping constant
H_{eff}	Effective field
H^e, H_k	External applied field and shape anisotropy field
M	Magnetization
M_s	Saturation magnetization of material
m, h	Normalized magnetization and effective magnetic field
$coeff$	Coefficient between read-current and produced external magnetic field
σ_H, V_H	Quantum Hall conductance and quantum Hall voltage

A typical TI-based memory device is built by a two-layer structure with ferromagnetic layer on the top and topological insulator layer on the bottom, as shown in Figure 2.5a. The TI device

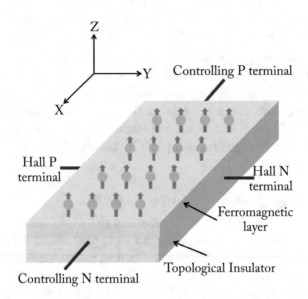

(a)

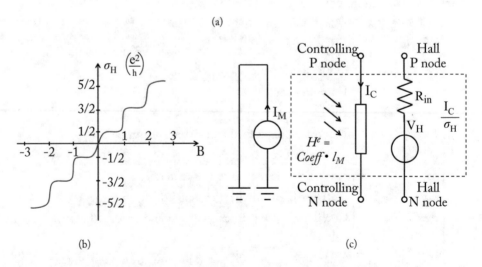

(b)

(c)

Figure 2.5: (a) Device structure; (b) schematic diagram of quantum Hall conductance; and (c) abstracted equivalent device circuit model.

has four terminals, with two controlling terminals along x-axis and two Hall terminals attached to the lateral sides. As discussed in Fujita et al. [2011], one bit can be then stored by the perpendicular magnetization of the ferromagnetic layer. Programming a bit requires an external magnetic field whose field strength is exceeding the coercivity of ferromagnetic layer. Under the magnetic field of the ferromagnetic layer, the topological insulator exhibits a quantum Hall conductance, as shown in Figure 2.5b. It can be observed that the sign of quantum Hall conductance is determined by the magnetic field orientation, thus the stored bit can be read out by detecting the sign of quantum Hall voltage

$$V_H = \frac{I_R}{\sigma_H},$$ (2.16)

where I_R is the applied read-current pulse along x axis.

The Hall conductance σ_H can be calculated by [Qi et al., 2006]

$$\sigma_H = \frac{e^2}{\hbar} \int \frac{d^2\vec{k}}{(2\pi)^2} (f_c - f_v)(\vec{k})\Omega_z(\vec{k}),$$ (2.17)

where the $\vec{k}$ is the wave vector, Ω_z the Berry curvature, and f_c and f_v the Fermi-Dirac distributions of conduction band and valence band, respectively.

It can be seen that the Hall conductance σ_H is a function of band-gap, Fermi Level and temperature. When the band-gap $\Delta \gg k_B T$, Equation 2.17 becomes

$$\sigma_H \approx \frac{e^2}{2h} sgn(M).$$ (2.18)

The h here is the Planck constant, e is the charge of electron, and $sgn(M)$ is the orientation of magnetization. As such, the quantum Hall conductance is a constant approximately equals to $19.4\mu S$. Note that the TI device is insensitive to disorder, imperfection, and cell geometry, which ensures a constant readout voltage even in the presence of perturbations.

From Equation 2.18, the quantum Hall voltage can be regarded as a current-controlled voltage source with the coefficient of σ_H, as shown in Figure 2.5c. Note that this equivalent quantum Hall voltage source has very limited driving ability, and a large internal resistance R_{in} is introduced for an accurate modeling.

More importantly, in order to model the dynamic behavior of the TI device during programming procedure, the magnetization trajectory needs to be studied, which is described by the normalized Landau-Lifshiltz-Gilbert equation (LLG) at macro-scale

$$\frac{dm}{d\tau} = -m \times h + \alpha \left(m \times \frac{dm}{d\tau} \right),$$ (2.19)

where the normalized effective field h equals to $-\frac{\delta\epsilon}{\delta m}$.

The ϵ is the normalized energy density:

$$\epsilon = \frac{1}{2} h_k (1 - m_z^2) - m \cdot h^e,$$ (2.20)

where the two energy density contributions are associated with anisotropy field and the external field, respectively.

Note that the required external magnetic field for device programming is generated by a current, as shown in Figure 2.5c. There exists a coefficient between the read-current and the produced field,

$$H^e = coeff \cdot I_M. \tag{2.21}$$

We assume the external magnetic field only have a perpendicular component along the easy axis z. Thus, normalized effective field $h = coeff \cdot I_M/M_s + h_k \cdot m_z$ can be obtained.

The solution of Equation 2.19 can be interpreted as the change of the normalized magnetization (m) over time. The normalized magnetization m can be expressed in spherical coordinates with variables θ and ϕ, as shown in Figure 2.2. The dynamic behavior described by θ and ϕ can be finally calculated by LLG

$$\theta = \theta_0 Exp\left(-\frac{t}{t_0}\right) \cdot \cos(\phi) \tag{2.22}$$

$$\omega = \frac{d\phi}{dt} = k_c \sqrt{k_d - \left(\alpha h_k - \alpha h_x^e\right)^2}, \tag{2.23}$$

where θ_0 is the initial value of θ, slightly tilted from the stable z or $-z$ directions, t_0 is procession time constant, and ω is the angular speed of θ; $k_c = \gamma_0 \cdot M_s$ is product of gyro-magnetic ratio and saturation magnetization; and $k_d \approx \frac{H_k}{M_s}$. Definitions of the remaining variables are also shown in Table 2.4.

The new state variable vector $X = [v_n, j_i, j_l, s_m]^T$ contains nodal voltage, source current, inductor current, and the new state variable s_m (magnetization angles θ and ϕ) for a TI device, respectively. We assume that (i) the conductance along x axis shows only weak dependency on the new state variables θ and ϕ; and (ii) the magnetization is only subject to the external field, and hence $S \approx 0$, $K_v^f \approx 0$ and $K_v^g \approx 0$.

As a result, according to Equation 2.10, we can obtain the required Jacobian terms as follows:

$$G = \begin{bmatrix} \sigma_x & -\sigma_x \\ -\sigma_x & \sigma_x \end{bmatrix};$$

$$K_s^f = \begin{bmatrix} 1 & -\frac{df(\phi_m,t)}{d\phi_m} \\ 0 & 0 \end{bmatrix}; \quad K_s^g = \begin{bmatrix} 0 & 0 \\ 1 & 0 \end{bmatrix},$$

where $f(\phi_m, t)$ is the right-hand-side of Equation 2.22, and σ_x is the conductance along the x-axis provided as a model parameter. Due to the non-scattering property of TI surface, an extremely high σ_x can be expected, which will contribute to an ultra-low power consumption.

Besides the magnetization dynamics, the derived MNA also has to model the quantum Hall voltage readout behavior. As discussed in the last subsection, the quantum Hall voltage can be modeled as a current-controlled voltage source, with coefficient as $\frac{1}{\sigma_H}$. As such, applying Equation 2.16, the incident matrix for the quantum Hall voltage source can be obtained as

$$E_i = \begin{bmatrix} 1 \\ -1 \end{bmatrix}.$$

Meanwhile, I_c/σ_H has to be added to the corresponding right-hand-side of Equation 2.10, where I_c is the read-current flowing through x-axis, and σ_H can be calculated by Equation 2.17. Therefore, all Jacobian terms required for the linearized system Equation 2.10 can be established, with which the TI device simulation considering the new state variables can be implemented in the SPICE-like simulator accordingly.

Simulation Results

In this section, the design of TI device based non-volatile memory (NVM) is discussed. First, a memory cell circuit is proposed with addressability achieved. Then, a memory array design is further illustrated with word read and write operations.

Memory Cell Circuit Design

Inspired by the toggle MRAM design [Engel et al., 2005], the cell structure of a TI device for NVM application is proposed in Figure 2.6. To program a selected TI cell, both word-write-line (WWL) and bit-write-line (BWL) produce half of the required external magnetic field H_e.

In order to achieve addressability, the amplitude of H_e is subject to:

$$H_e/2 \leq H_c \leq H_e, \tag{2.24}$$

where the H_c is the coercivity of the magnetic surface. Here the currents along WWL and BWL to produce $H_e/2$ are defined as programming current I_{PW} and I_{PB}, respectively. The upward I_{PB} and leftward I_{PW} are defined as positive. If both I_{PW} and I_{PB} are applied, there exists an external magnetic field of H_e that exceeds the coercive field of ferromagnetic layer to program a cell. If I_{PW} is applied only or I_{PB} is applied only, the cell is exposed to a magnetic field of $H_e/2$, which is insufficient to switch the magnetization. A zero magnetic field, when no current is applied for both WWL and BWL, is also not able to switch the magnetization. Thus, the cell addressability can be achieved under programming operation. Moreover, to read a cell, the cell is first selected by the signal through word-line, and then a read-current pulse I_R is applied through bit-line. The corresponding quantum Hall voltage V_H will be either positive or negative depending on the status of magnetization orientation. Therefore, by comparing V_H with reference voltage, the bit stored by magnetization orientation can be known.

Memory Array Circuit Design

Figure 2.7 shows a 4 × 4 memory array design based on the TI memory cell. It can be observed that it is impossible to write 1-bits and 0-bits for a word at one time since they require opposite

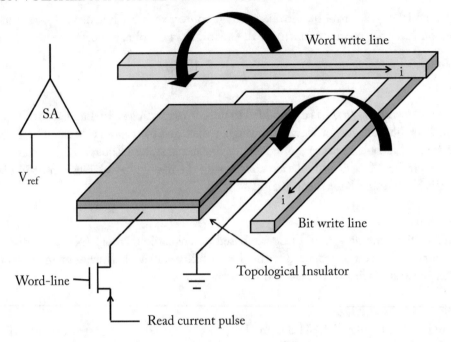

Figure 2.6: Cell circuit of topological insulator based memory.

I_{PW} and I_{PB} directions. Thus to write a word, writing 1-bits or 0-bits is conducted separately. To write 1001 for *word*0 for instance, WWL0, BWL0 and BWL3 are applied with positive I_{PW} and I_{PB} while other WWLs and BWLs are applied with no current. As discussed in the last subsection, the addressability allows only *bit*0 and *bit*3 to be programmed into 1, while other bits remain their status. Then WWL0, BWL1, and BWL2 are applied with negative I_{PW} and I_{PB} so that *bit*1 and *bit*2 are programmed into 0. As such, 1001 is successfully written to *word*0 in two separate steps. Because there exists an inversely proportional relationship between current induced magnetic field and distance in space, and also due to the programming threshold for magnetic field, the operations on target cells will not interfere their neighboring cells. To read out a word, its corresponding word-line is selected, and a current pulse I_R is applied for all bit-lines. As a result, the V_H for each bit can be interpreted by sense amplifiers for each bit-line.

A TI-based NVM design platform is developed. As CMOS circuits are still required as interfacing part, hybrid CMOS-TI simulation is required. Similar to a BSIM model for MOS-FET, the physical model of TI device is implemented into a SPICE-like simulator NGspice [Nenzi and Holger, 2010]. Based on the developed SPICE-like simulator, a number of experiments have been conducted for TI-based NVM designs. In the following numerical experiments, we first validate our physical model against the reported measurement data [Koch et al., 1998]. Then, we design the cell and array memory circuits of TI and verify them by the SPICE-like

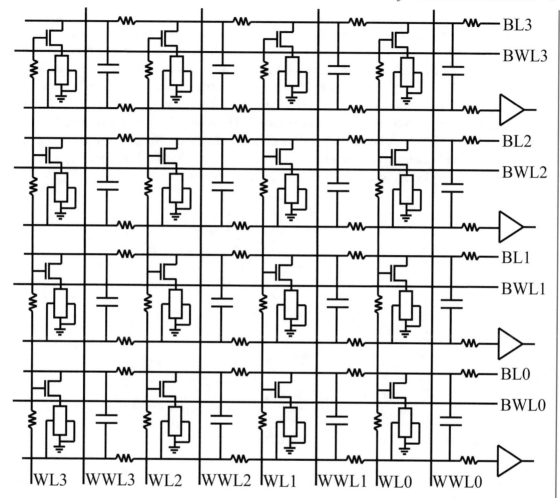

Figure 2.7: One 4 × 4 topological insulator-based NVM array.

simulator under both read and write operations. Finally, we compare the performance of read and write power with the other emerging NVM devices. All numerical experiments are implemented in C and are conducted on the same work-station with Intel Core i5 CPU and 8G RAM.

Validation of Dynamic Effect with New State Variable of TI Device

In order to validate the dynamic model of magnetization for TI device, we show the device level simulation results based on soft ferromagnet material $Ni_{60}Fe_{40}$ with parameters extracted from measurement in Koch et al. [1998]. The lateral dimension of TI device is 0.8 by 1.6 μm^2, with ferromagnet layer thickness of 5 nm. The saturation magnetization is set to 740 kA/M, the shape anisotropy is field 1.72 kA/M, the damping constant is 0.01, and the current-to-magnet coefficient is 10^6. Same parameters are assumed in the following experiments for consistency.

Figure 2.8 shows the magnetization switching time vs. the applied external magnetic field amplitude. We can see the results produced by our simulator fit well with the measured data reported in Koch et al. [1998]. A nearly inversely proportional relationship can be observed between switching time and magnetic field. For example, in order to achieve a faster memory programming speed, a stronger magnetic field is desirable. Moreover, a switching threshold can be observed from Figure 2.8. So only when the external magnetic field exceeds a certain strength, approximate 2 mT observed from figure, can the magnetization be switched. This is also consistent with Koch et al. [1998], where the magnetic coercivity is reported as 2 mT.

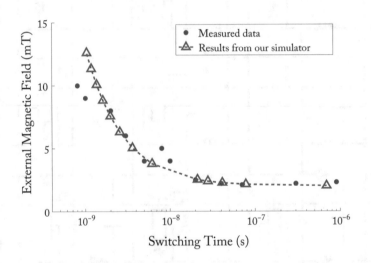

Figure 2.8: Validation of the switching time and external magnetic field relationship for magnetization dynamics.

As such, the magnetization dynamics has been verified validate for our developed TI device simulator. Note that for TI with other materials as ferromagnetic layer, parameters of saturation magnetization, shape anisotropy field, and damping constant need to be specified for the simulator in a model file.

Hybrid Simulation with CMOS for TI-based Memory Cell

In order to investigate the performance of TI device-based memory, transient analysis is conducted for one TI memory cell circuit as illustrated in Section 2.2.3. The technology node is 65 nm for CMOS part, and VDD is 1.2 V. The aspect ratio of transistors are set to 3. The read-current pulse is set to be 1 μA with a pulse width of 5 ns. The programming current of both word-write-line I_{PW} and bit-write-line I_{PB} are generated by their respective current sources. As discussed previously, strong external magnetic field is desirable for a fast programming speed. Moreover, in order to achieve memory cell addressability, however, the field strength should be subject to Equation 2.24. So in this work, the magnetic coercivity H_c is 2 mT, and H_e is designed at 3 mT. Parameters for ferromagnetic layer are the same as last section.

Figure 2.9 shows the dynamic response when using the new state variables. Both programming current I_{PW} and I_{PB} are applied at zero second. It can be observed that θ starts to deviate from the original angle once H_e is applied. Its maximum deviation increases exponentially with time before the reversal happens. The fluctuation decays very fast after reversal since the shape anisotropy field alters its direction to strengthen H_e, and then the device enters into the other stable state. The whole process indicates that a write latency, i.e., switching time, of about 10 ns is achieved under 3 mT magnetic field. The continuously increasing ϕ shows a very fast circulating frequency in the x-y plane, which causes θ to fluctuate in the same frequency.

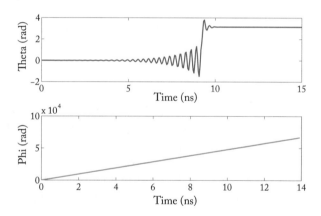

Figure 2.9: Dynamic response of topological insulator new state variables.

The simulation result for the cell circuit readout operation is shown in Figure 2.10. First, the word-line signal is set to logic-1 before the readout operation. Then at the time of 12 ns, a read-current pulse with amplitude of 1 μA is applied through bit-line. The quantum Hall voltage responds simultaneously, and is fed into the sense amplifier (SA) later. After a time of T_{delay}, which is the combination of bit-line delay and SA sensing delay, the SA outputs the stored logic. Note that the T_{delay} depends on the array size and quantum Hall voltage amplitude in practice, and the read-current pulse width is set to 5 ns to secure a successful readout operation.

Performance Comparison of TI-based Memory Array
Here we further investigate the TI-based memory array design, as shown in Figure 2.7. The circuit settings are same with Section 2.2.3. The transient analysis conducted is to write 1001 to *word0*

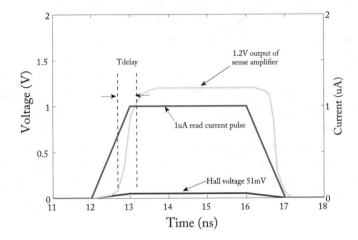

Figure 2.10: Read operation for topological insulator based memory cell.

as discussed. The *word*0 is initialized 0110 for better illustration, the other bits are initialized all zeros.

Figure 2.11 shows the timing diagram of write and read operations. Note that the positive and negative current are indicated by 1 and −1 in this figure. It can be observed that the write operation is executed in two phases. First, the first bit and fourth bit of *word*0 are written to 1 in the first 10 ns. From 10 ns to 20 ns, the second and third bits are written to 0. To read out *word*0, *word*0 is selected through word-line, and bit-lines of all bits are applied with read current pulse. A correct readout can be observed. It has also been verified that all the other bits remain their initial values.

Table 2.5 shows the comparison of performance for different memory technologies. Compared with the other emerging NVMs, such as phase change memory (PCM), spin-transfer torque magnetic tunneling junction (STT-MTJ), and ReRAM, the TI device-based memory shows both faster read and write latencies. It is also noticed that the TI-based memory exhibits in several orders of magnitude lower write energy. The write energy of TI is calculated by Equation 2.20 with device dimension. Actually, the read energy of TI is also extremely low due to the non-scattering property. Simulation shows a read energy of 1.2e−17J/bit. The data for other memory technologies is extracted from ITRS 2011 [ITRS, 2010].

2.3   DOMAIN WALL DEVICE AND MODEL

Domain wall nanowire, also known as racetrack memory [Parkin et al., 2008, Thomas et al., 2011, Venkatesan et al., 2012], is a newly introduced non-volatile memory device in which multiple bits of information are stored in single ferromagnetic nanowire. As shown in Figure 2.12a, each bit

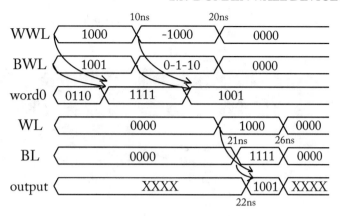

Figure 2.11: Timing diagram for word write and read operations in a 4×4 TI array.

Table 2.5: Performance comparison for different memory technologies

Memory Technology	Write Latency (ns)	Read Latency (ns)	Write Energy (J/bit)
SRAM	0.2	0.2	5e-16
DRAM	2-10	2-10	4e-15
PCM	100	12	6e-12
STT-MTJ	35	35	2.5e-12
FeRAM	40	60	3e-14
TI	20	5	1e-17

is denoted by the leftward or rightward magnetization direction, and adjacent bits are separated by domain walls. By applying a current through the shift port at the two ends of the nanowire, all the domain walls will move left or right at the same velocity while the domain width of each bit remains unchanged, thus the stored information is preserved. Such a tape-like operation will shift all the bits similarly like a shift register.

In order to access the information stored in the domains, a strongly magnetized ferromagnetic layer is placed at a desired position of the ferromagnetic nanowire and is separated by an insulator layer. Such a sandwich-like structure forms a magnetic-tunnel-junction (MTJ), through which the stored information can be accessed. In the following, the write, read, and shift operations are modeled, respectively.

2.3.1   MAGNETIZATION REVERSAL

The write access can be modeled as the magnetization reversal of MTJ free layer, i.e., the target domain of the nanowire. Note that the dynamics of magnetization reversal can be described by

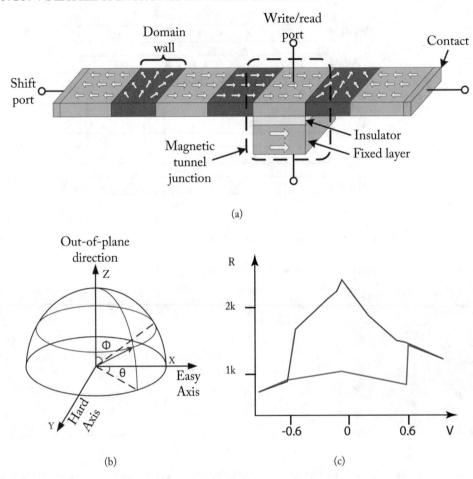

Figure 2.12: (a) Schematic of domain wall nanowire structure with access port and shift port; (b) magnetization of free-layer in spherical coordinates with defined magnetization angles; and (c) typical R-V curve for MTJ.

the precession of normalized magnetization m, or state variables θ and ϕ in spherical coordinates, as shown in Figure 2.12b. The spin-current induced magnetization dynamics described by θ and ϕ is given by

$$\theta = \theta_0 Exp\left(-\frac{t}{t_0}\right) \cdot \cos(\phi) \tag{2.25}$$

$$\omega = \frac{d\phi}{dt} = k_1\sqrt{k_2 - (k_3 - k_4 I)^2}, \tag{2.26}$$

where θ_0 is the initial value of θ, slightly tilted from the stable x or $-x$ directions; t_0 is procession time constant; ω is the angular speed of ϕ; k_1 to k_4 are magnetic parameters with detailed explanation in Chapter 2; and I is the spin-current that causes the magnetization precession.

2.3.2   MTJ RESISTANCE

A typical R-V curve for MTJ is shown in Figure 2.12c with two regions: giant magnetoresistance (GMR) region and tunneling region. Depending on the alignment of magnetization directions of the fixed layer and free layer, parallel or anti-parallel, the MTJ exhibits two resistance values R_l and R_h. As such, the general MTJ resistance can be calculated by the giant magnetoresistance (GMR) effect

$$R(\theta_u, \theta_b) = R_{l0} + \frac{R_{h0} - R_{l0}}{2}(1 - \cos(\theta_u - \theta_b)), \tag{2.27}$$

where θ_u and θ_b are the magnetization angles of upper free layer and bottom fixed layer, R_{l0} and R_{h0} are the MTJ resistances when the applied voltage is subtle. When the applied voltage increases, there exists tunneling effect caused voltage-dependent resistance roll-off,

$$\begin{cases} R_l(V) = \dfrac{R_{l0}}{1 + c_l V^2} \\ R_h(V) = \dfrac{R_{h0}}{1 + c_h V^2}, \end{cases} \tag{2.28}$$

where c_l and c_h are voltage-dependent coefficients for parallel state and anti-parallel state, respectively.

2.3.3   DOMAIN WALL PROPAGATION

Like a shift register, the domain wall nanowire shifts in a digital manner, thus could be digitalized and modeled in the unit of domains, in which a bit is stored. Note that except the bit in the MTJ, the other bits denoted by the magnetization directions are only affected by their adjacent bits. In other words, the magnetization of each bit is controlled by the magnetization in adjacent domains. Inspired by this, we present a magnetization controlled magnetization (MCM) devices based behavioral model for domain wall nanowires. Unlike the current-controlled and voltage-controlled devices, the control in MCM device needs to be triggered by rising edge of one SHF-signal, which can be formulated as

$$\begin{aligned} \theta &= f(T_{sl}, \theta_r, T_{sr}, \theta_l, \theta_c) \\ &= T_{sl}\theta_r + T_{sr}\theta_l + \overline{T}_{sl}\overline{T}_{sr}\theta_0. \end{aligned} \tag{2.29}$$

in which T_{sl} and T_{sr} are the shift-left and shift-right commands; θ_r and θ_l are the magnetization angles in right adjacent cell and left adjacent cell, respectively; and θ_c is the current state before the trigger signal. This describes that the θ-state will change when triggered and will remain state if no shift-signal is issued.

For the bit in MTJ, the applied voltage for spin-based read and write will also determine the θ-state as discussed previously. Therefore we have,

$$\theta = f(T_{sl}, \theta_r, T_{sr}, \theta_l, \theta_0) + g(Vp, Vn, \theta_c), \tag{2.30}$$

where Vp and Vn are the MTJ positive and negative nodal voltages, and $g(Vp, Vn, \theta_0)$ is the additional term that combines Equation 2.25–2.28.

In addition, the domain wall propagation velocity can be mimicked by the SHF-request frequency. The link between the SHF-request frequency and the propagation velocity is experimentally observed by current-velocity relation [Chiba et al., 2010],

$$v = k(J - J_0), \tag{2.31}$$

where J is the injected current density and J_0 is the critical current density.

By combining Equation 2.22–2.30, with the magnetization angles θ and ϕ as non-volatile state variables other than electrical voltages and currents, one can fully describe the behaviors of the domain wall nanowire device, where each domain is modeled as the proposed MCM device. As such, the modified nodal analysis (MNA) can be built in NVM-SPICE to verify circuit designs by domain wall nanowire devices.

2.3.4   CIRCULAR DOMAIN WALL NANOWIRE

In traditional domain wall nanowire, energy consumed by write and shift operations, even though minimized, makes it impossible to be energy-level comparable with CMOS. In order to improve the energy-performance further, two key observations are found in the logic operations analysis.

- Many of the write-operations, and hence, the subsequent shift-operations are not necessary since the previous and new outputs are same. Energy can be reduced greatly if these redundant operations are prevented.

- Write energy is larger than domain wall shift energy. As in Fukami et al. [2009], the write operations can be replaced with shift operations.

For subsequent optimization, a scheme is proposed to include both these observations. In this scheme, XOR operation is utilized to detect cases where the new output matches the previous output. So, if the Boolean logic performed under the read circuit is denoted by f' and its output is C, the operation performed under the read circuit will be denoted by $f = f \oplus C$. The output of the read circuit will be 1 if the new output is different from the old value, else 0. So, if $f = AB$, the operation performed under the read circuit is $f = AB \oplus C$. Instead of switching the output of an MTJ to store the output, domain wall shifting can be used to store the output. Since the output of the read operation is either 1 or 0, a circular ferromagnetic nanowire can be used to store the correct output. The rationale behind choosing a circular racetrack can be clearly understood as we describe its operation.

Figure 2.13 shows a ring-shaped circular domain wall nanowire. The ring has 4 pinning sites—P1 to P4, and two domain walls—DW_A and DW_B. As we know, a domain wall exists at the boundary of 1 and 0. Two pairs of metal contacts are placed on the ring. Current pulses are injected into the ring for STT-driven motion of the domain walls. C and D represent the output of the read operation as mentioned above, and D is to be used as input by other logic gates. MTJs are placed on the ring to sense the values of $C, \overline{C}, D,$ and $\overline{D}$. Initially, the two domain walls are pinned at P1 and P3.

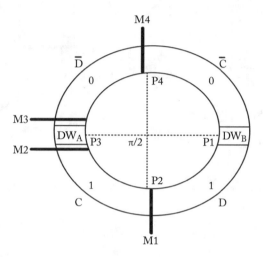

Figure 2.13: Circular domain wall nanowire.

If the output of the read operation is 0, the new output is the same as the current value of $C(= 1)$ and so no shifting of domain walls occurs. If the output of the read operation is 0, the positions of A and B are interchanged, so that the value of C and D. The sequence of steps for interchanging the domain wall positions are as follows.

- Figure 2.14a: At time $t = 0$, M1 is grounded and a current pulse is applied at M2. M3 and M4 are disconnected from ground or current source by switching OFF their pass transistors. Current density is greater than the critical value for DW propagation. The width (δ) and direction of the current pulse is such that A and B travel to P4 and P2, respectively.

- Figure 2.14b: At $t = \triangle$, M3 is grounded and a current pulse is applied at M4. M1 and M4 are disconnected from ground or current source. A and B travel to P1 and P3, respectively.

- Figure 2.14c: Thus, the values of C and D are reversed at $t = 2\triangle$.

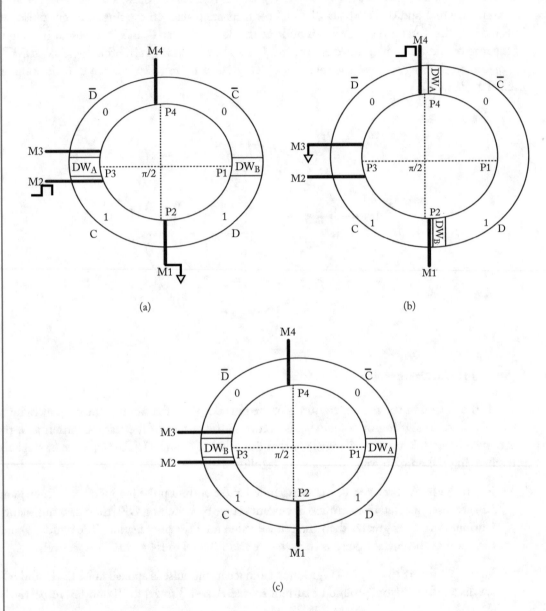

Figure 2.14: Position of domain walls at (a) $t = 0$; (b) $t = \Delta$; (c) $t = 2\Delta$.

2.4 SPINTRONIC STORAGE

2.4.1 SPINTRONIC MEMORY

Spintronic memory is similar to conventional SRAM/DRAM technologies which require transistors to control. Thus, spintronic memory is often associated with 1T-1R structure, where "T" stands for transistor and "R" denotes one non-volatile device whose state is represented by resistance.

A typical hybrid STT-RAM cell is shown in Figure 2.15a with one transistor and one STT-MTJ in series connection.

The structure is identical to that of DRAM cell except that the capacitor is replaced by STT-MTJ device. The gate of transistor is connected to word-line, which serves to select target cells in same word-line. When enabled, two bit-lines (named bit-line and select line to distinguish) can be driven to have V_w or $-V_w$ depending on the desired data to write, 1 or 0. In the "write 1" operation, WL is connected to VDD, and BL and SL are connected to V_{DD} and ground, respectively. In the "write 0" operation the polarities of SL and BL line are interchanged. Readout operation can be performed in similar way by using V_r, and the current loop will eventually be measured by readout circuit, which determines device state by current amplitude.

2.4.2 SPINTRONIC READOUT

Existing STT-RAM readout schemes to avoid disturbance of large STT-MTJ resistance variation usually require several steps, which slows down the read latency. We show that by applying a single-sawtooth pulse and exploiting the resistance roll-off of STT-MTJ, the robust readout can be achieved within one cycle.

Basic STT-RAM Readout Circuit

The basic voltage-sensing scheme for the popular 1T-1MTJ structure STT-RAM is shown in Figure 2.16a. The reference voltage is set to satisfy

$$I_r \cdot (R_{AP} + R_t) > V_{ref} > I_r \cdot (R_P + R_t),$$

where I_r is the applied read current, R_{AP}, R_P, and R_t are the MTJ anti-parallel state resistance, parallel state resistance, and cell transistor ON-state r_{ds}, respectively. However, in the presence of bit-to-bit MTJ resistance variation, the reference voltage has to fulfill

$$Min(V_{BL,AP}) > V_{ref} > Max(V_{BL,P}),$$

where a satisfying V_{ref} may not exist when variation is large.

Destructive Self-reference Readout Circuit

In order to achieve reliable readout in the presence of large MTJ resistance variation, a self-reference readout is presented in Jeong et al. [2003], whose diagram is shown in Figure 2.16b. The read-operation is done in five phases:

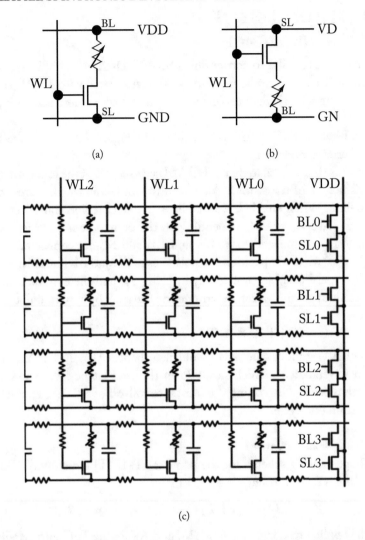

Figure 2.15: Circuit diagrams of 1T-1R STT-RAM memory cell: (a) simplified memory cell for write 1; (b) simplified memory cell for write 0; and (c) 16-bit STT-RAM with 4 bit-lines and 4 word-lines.

- the read-current I_r is applied and its bit-line voltage is stored in C1;

- the "0" (parallel state) value is written to the target cell;

- the read-current I_r is applied again and its bit-line voltage is stored in C2;

- the sense amplifier is enabled and voltages of C1 and C2 are compared; and the output is "1" (anti-parallel state) if V_{C1} is greater than V_{C2} and "0" otherwise; and

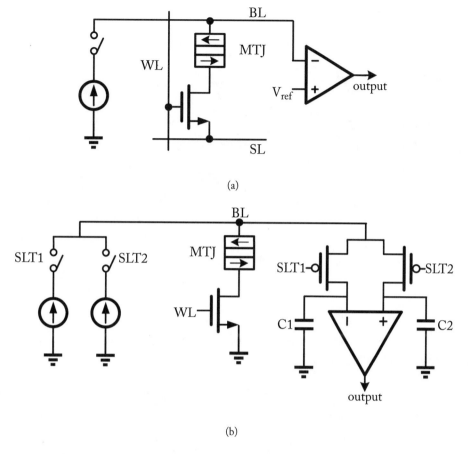

Figure 2.16: The existing schemes for STT-RAM readout: (a) basic STT-RAM readout; (b) destructive self-reference readout in Jeong et al. [2003]. *(Continues.)*

- the output value has to be written back to the destructed cell.

Therefore, in terms of both speed and power, the overhead brought by write-back may be large for this scheme.

Non-destructive Self-reference Readout Circuit

The current-dependent resistance roll-off can be observed for STT-MTJ, as shown in Figure 2.17. By exploiting the fact that the roll-off slope of the anti-parallel state is much greater than that of parallel state, a non-destructive self-reference readout is proposed in Chen et al. [2010] with a diagram shown in Figure 2.16c.

The read-operation is done in three phases:

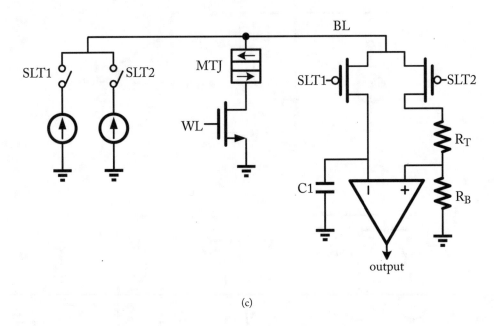

(c)

Figure 2.16: *(Continued.)* The existing schemes for STT-RAM readout: (c) non-destructive self-reference readout in Chen et al. [2010].

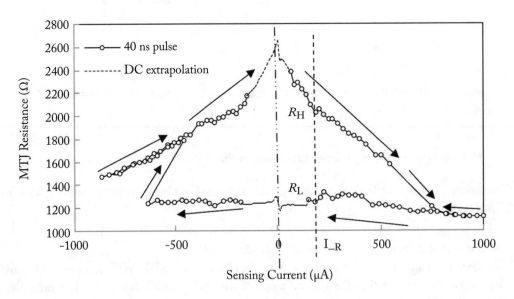

Figure 2.17: The measured R-I sweep curve of a typical MgO-based MTJ in Jeong et al. [2003].

- the read-current I_{r1} is applied to achieve its corresponding resistance R_1;

- the read-current I_{r2} is applied to achieve its corresponding resistance R_2; and

- the sense amplifier is enabled and R_1 and R_2 are compared. The output is "1" if two values are significantly different and "0" otherwise.

As such, the non-destructive self-reference readout can improve the read latency by eliminating the two time-consuming write phases. However, this scheme still has limited performance in read latency and sensing margin, as discussed below.

2.5   SPINTRONIC LOGIC

2.5.1   XOR

The magnetization switching with sub-nanosecond speed and and sub-pJ energy have been experimentally demonstrated [Rowlands et al., 2011, Zhao et al., 2011, 2012]. As such, the domain wall nanowire based logic can be further explored for logic-in-memory based computing. In this section, we show how to further build DWL-based XOR-logic, and how it is applied for low-power ALU design for comparison and addition operations.

The GMR-effect can be interpreted as the bitwise-XOR operation of the magnetization directions of two thin magnetic layers, where the output is denoted by high or low resistance. In a GMR-based MTJ structure, however, the XOR-logic will fail as there is only one operand as variable since the magnetization in fixed layer is constant. Nevertheless, this problem can be overcome by the unique domain wall shift-operation in the domain wall nanowire device, which enables the possibility of DWL-based XOR-logic for computing.

A bitwise-XOR logic implemented by two domain wall nanowires is shown in Figure 2.18. The proposed bitwise-XOR logic is performed by constructing a new read-only-port, where two free layers and one insulator layer are stacked. The two free layers are in the size of one magnetization domain and are from two respective nanowires. Thus, the two operands, denoted as the magnetization direction in free layer, can both be variables with values assigned through the MTJ of the according nanowire. As such, it can be shifted to the operating port such that the XOR-logic is performed.

For example, the $A \oplus B$ can be executed in the following steps.

- The operands A and B are loaded into two nanowires by enabling WL_1 and WL_2, respectively;

- A and B are shifted from their access-ports to the read-only-ports by enabling SHF_1 and SHF_2, respectively;

- By enabling RD, the bitwise-XOR result can be obtained through the GMR-effect.

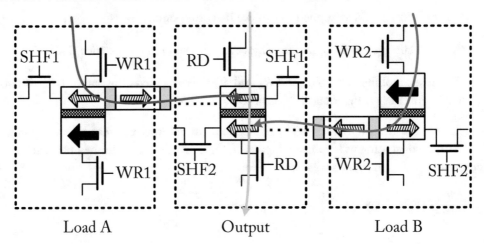

Figure 2.18: Low-power XOR-logic implemented by two domain wall nanowires.

Note that in the x86 architecture processors, most XOR instructions also need a few cycles to load its operands before the logic is performed, unless the two operands are both in registers. As such, the proposed DWL-based XOR-logic can be a potential substitution of the CMOS-based XOR-logic. Moreover, similar as the DWM macro-cell, zero leakage can be achieved for such XOR-logic.

The transient analysis of the domain wall nanowire XOR structure has been performed in the SPICE simulator with both the controlling timing diagram and operation details shown in Figure 2.19.

The current density of 7e10 A/m^2 is utilized for magnetization switching. The θ states of the nanowire that takes A are all initialized at 0, and the one takes B all at π. Only two-bit per nanowire is assumed for both nanowires. The operating-port is implemented as a developed magnetization controlled magnetization (MCM) device, with internal state variables θ and ϕ for both upper layer and bottom layer. In the cycles of *load A* and *load B*, the precession switching can be observed for the MTJs of both nanowires. Also, the switching energy and time have been calculated as 0.27 pJ and 600 ps, which is consistent with the reported devices [Rowlands et al., 2011, Zhao et al., 2011, 2012]. In the *shift* cycles, triggered by the *SHF*-control signal, the dynamics θ and ϕ of both upper and bottom layers are updated immediately. In the *operation* cycle, a subtle sensing current is applied to provoke GMR-effect. Subtle magnetization disturbance is also observed in both layers in the MCM device, which validates the read-operation. The θ values that differ from initial values in the *operation* cycle also validate the successful domain wall shift.

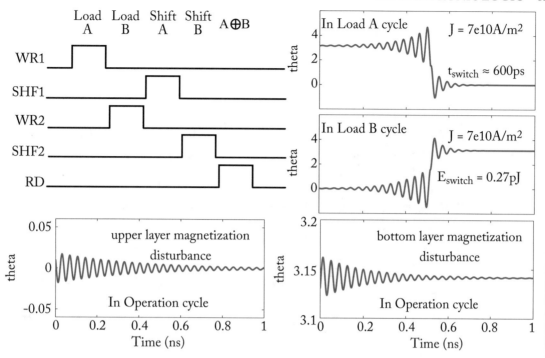

Figure 2.19: The timing diagram of DWL-XOR with SPICE-level simulation for each operation.

2.5.2 ADDER

To realize a full adder, one needs both *sum* logic and *carry* logic. As the domain wall nanowire-based XOR logic has been achieved, the sum logic can be readily realized by deploying two units: $Sum = (A \oplus B) \oplus C$. As for carry logic, a spintronics-based carry operation is proposed in Trinh et al. [2012], where a pre-charge sensing amplifier (PCSA) is used for resistance comparison. The carry logic by PCSA and two branches of domain wall nanowires is shown in Figure 2.20a. The three operands for carry operation are denoted by resistance of MTJ (low for 0 and high for 1), and belong to respective domain wall nanowires in the left branch. The right branch is made complementary to the left one. Note that the C_{out} and $\overline{C_{out}}$ will be pre-charged high at first when PCSA EN signal is low. When the circuit is enabled, the branch with lower resistance will discharge its output to "0". For example, when left branch has no or only one MTJ in high resistance, i.e., no carry out, the right branch will have three or two MTJs in high resistance, such that the C_{out} will be 0. The complete truth table is shown in Figure 2.20b, which is able to confirm carry logic by this circuit. The domain wall nanowire works as the writing circuit for the operands by writing values at one end and shift it to PCSA. Note that with the full adder implemented by domain wall nanowires and intrinsic shift ability of domain wall nanowire, a shift/add multiplier can be further achieved purely by domain wall nanowires.

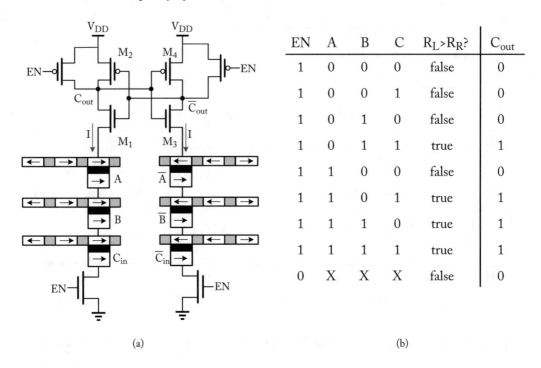

EN	A	B	C	$R_L > R_R$?	C_{out}
1	0	0	0	false	0
1	0	0	1	false	0
1	0	1	0	false	0
1	0	1	1	true	1
1	1	0	0	false	0
1	1	0	1	true	1
1	1	1	0	true	1
1	1	1	1	true	1
0	X	X	X	false	0

(a)   (b)

Figure 2.20: The carry-out logic achieved by domain wall nanowires.

2.5.3 MULTIPLIER

With the full adder implemented by domain wall nanowires and intrinsic shift ability of domain wall nanowire, the multiplication operation can be easily achieved by breaking it down to multiple domain wall shift operations and additions. Operand A with m non-zero bits multiplied by operand B with n non-zero bits ($m > n$) can be decomposed into n shift operations and n additions. For example, multiplication of binary 1011 and 110 can be decomposed into addition of 10110 and 101100, where 10110 and 101100 are obtained by left-shifting 1011 one and two bits in domain wall nanowire. As such, not only can the complicated domain wall multiplier circuit be avoided, but also multiplication operation can be handled more efficiently by reusing domain wall adders in a distributed MapReduce fashion.

2.5.4 LUT

Figure 2.21a shows the structure of cell in the LUT array. The access-port lies in the middle of the nanowire, which divides the nanowire into two segments. The left-half segment of nanowire is used for data storage while the right-half segment is reserved for shift-operation in order to avoid information lost.

Figure 2.21b shows the domain wall nanowire-based LUT array. The input of the function implemented by LUT is represented as binary address. The address is fed into word-line decoder and bit-line MUX to find the target domain wall nanowire cell, where the multiple-bit result is kept. The LUT array size depends on the domain, range, and precision of the function to perform.

Based on the way data is organized, the result can be output in serial manner or parallel manner. In serial output scenario, the binary result is stored in single domain wall nanowire that is able to hold multiple bits of information. Assume each cell has only one access port and the first bit of result is initially aligned with access port, the way to output result is to iteratively readout and shift one bit until the last bit is output. In parallel output scenario, the multiple-bit result is distributed into different nanowires. Because each cell has their own access port, the multiple bits can be output concurrently. The design complexity of parallel output scheme is that, to find the relative position of the result within the nanowire, a variable access time will be introduced. For example, if luckily the result is stored at first bit of the nanowires, the result can be readout in one cycle; on the contrary if the result is kept at the very last bit of the nanowires, it will take tens of cycles to shift first before the result is output. Therefore, the choice between serial output and parallel output is the tradeoff between access latency and design complexity.

Figure 2.22 shows the power characterization of DW-LUT in different array sizes. To obtain the area, power and speed of DW-LUT, the memory modeling tool CACTI [Thoziyoor et al., 2008] has been extended with domain wall nanowire model. In terms of dynamic energy per look-up operation, the parallel output scenario is much more power efficient than serial output scenario, and the gap enlarges when array size increases. This is because more cycles are required to output results in serial than in parallel, therefore more access operations are involved. However, the serial scenario is able to avoid the variable access latency issue, which reduces the design complexity of the controller. For leakage power, the non-volatility leads to extremely low leakage power in nW scale, which is negligible compared with its dynamic power. For volatile SRAM and DRAM, the leakage power may consume as large as half the total power especially in advanced technology node [Thoziyoor et al., 2008].

Once the domain, range, and precision of function are decided, the DW-LUT size can be determined accordingly. Therefore, the power characterization can be used as a quick reference to estimate the power profile of specific function to perform system level design exploration and performance evaluation.

2.6   SPINTRONIC INTERCONNECT

2.6.1   CODING-BASED INTERCONNECT

Designers of digital systems implement various low-power techniques at different abstraction levels to optimize their design for high energy efficiency. A significant portion of the energy

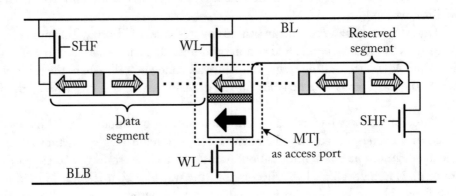

(a)

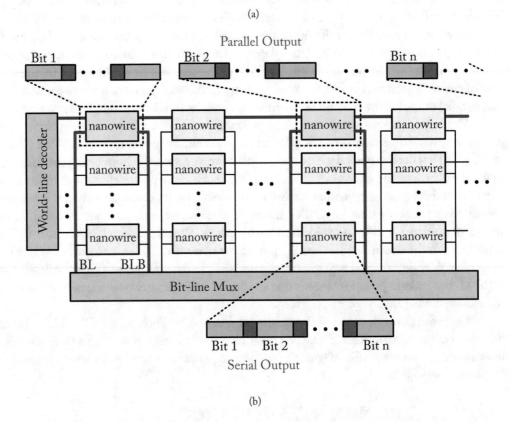

(b)

Figure 2.21: (a) Domain wall memory cell structure; and (b) LUT by domain wall nanowire array with parallel output and serial output.

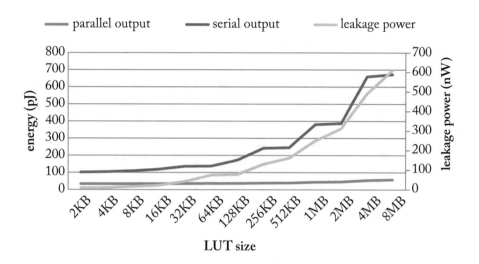

Figure 2.22: Power characterization for DW-LUT in different sizes.

consumed by digital systems comes from the switching activity of buses and interconnects [Burd and Peters, 1994, Sakurai, 2000]. Power crisis, interconnection crisis, and complexity crisis are described in Sakurai [2000]. For power crisis, some activities can lower the power consumption from device level to system level. For interconnection crisis, it includes cost, delay, power, and reliability. The complexity issue can only be solved by the sharing and re-use of design data.

Long length, large bit-width, and the large loads connected at the end of busses are responsible for their high capacitance [Liu and Svensson, 1994, Musoll et al., 1998, Sotiriadis and Chandrakasan, 2002]. High capacitance and large bus drivers make switching activity highly energy consuming in these interconnects. In Sotiriadis and Chandrakasan [2002], a comprehensive mathematical analysis of the energy dissipation in deep submicron technology buses is presented. The energy estimation is based on an elaborate bus model that includes distributed and lumped parasitic elements that appear as technology scale. Liu and Svensson [1994] developed a method and a tool for power modeling of CMOS VLSI chips, which makes it possible to estimate the power consumption of a chip based on gate count, memory size, logic, and layout styles. The tool is not as accurate as gate level simulators, but it gives a fast estimation far before circuit and layout design.

One of the ways to mitigate this issue is by reducing the number of bit-transitions in the interconnects. In this approach, the data is encoded as per a scheme before being transmitted over the bus or interconnect. The coding scheme is such that the encoded data transmit while incurring minimum switching activity in the interconnect. After the transmission is complete, the received data is decoded to get the original data. Figure 2.23 shows a simple block diagram to explain this technique. Numerous coding schemes have been proposed in literature to reduce these transitions in interconnects [Benini et al., 1997, 2000, Musoll et al., 1998, Ramprasad et al., 1999, Stan and Burleson, 1995]. Stan and Burleson [1995] proposed the Bus-Invert method of coding the I/O which lowers the bus activity and thus decreases the I/O peak power dissipation and the I/O average power dissipation. Ramprasad et al. [1999] presented a source-coding framework for the design of coding schemes to reduce transition activity. The schemes are suited for high-capacitance busses where the extra power dissipation due to the encoder and decoder circuitry is offset by the power savings at the bus.

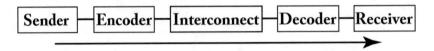

Figure 2.23: Coding-based interconnect architecture.

Works in Musoll et al. [1998] presents the working-zone encoding (WZE) method for encoding an external address bus, based on the conjecture that programs favor a few working zones of their address space at each instant to reduce the energy consumption due to input–output pins. As reported in Stan and Burleson [1997], a large portion of power dissipation in high performance CMOS VLSI is due to the inherent difficulties in global communication at high rates and they propose several approaches to address the problem. By carefully choosing the encoding, of these signals, the average and peak power dissipation can be minimized.

2.6.2   DOMAIN WALL-BASED ENCODER/DECODER

In our proposed design, we use Racetracks to implement such an encoder or decoder. Figure 2.24 shows an RM-based 2-bit encoder design. 2-bit data input to the design produces a 2-bit encoded output. The encoded bits for various inputs are stored in RM. Because it is a 2-bit design, 2 RMs are used for storing the encoded bits. Once the bits are stored in the RM, no power is needed to maintain the state of these bits in the RM. MTJ sensors placed on the RMs are used for reading the bits stored in the RMs. The state of these MTJs are sensed by Sensing Amplifiers (SAs) [Zhao et al., 2009].

The proposed encoder has two operation modes determined by the input Wr_{EN}. When $Wr_{EN} = 1$, output of gate $G1$ is 0 and so, the SAs are in pre-charge phase. Both the complementary outputs of the SAs are HIGH in this phase. When $Wr_{EN} = 1$ and the DW-shift clock $CLK_{sft} = 0$, the new input bits—$D0$ and $D1$—are written into the write-heads of the RMs. A 0 is written in the write-head when transistors $T1$ and $T3$ are in on-state. When $T2$ and $T4$ are ON, a 1 is written. When $CLK_{sft} = 1$, outputs of $G3$–$G6$ are 0. As a result, nMOS transistors $T3$, $T4$ of the write circuits are switched off, thereby, turning off the switching current through the write-heads of the RMs. But, the output of $G2$ is HIGH. This switches on the pass transistors

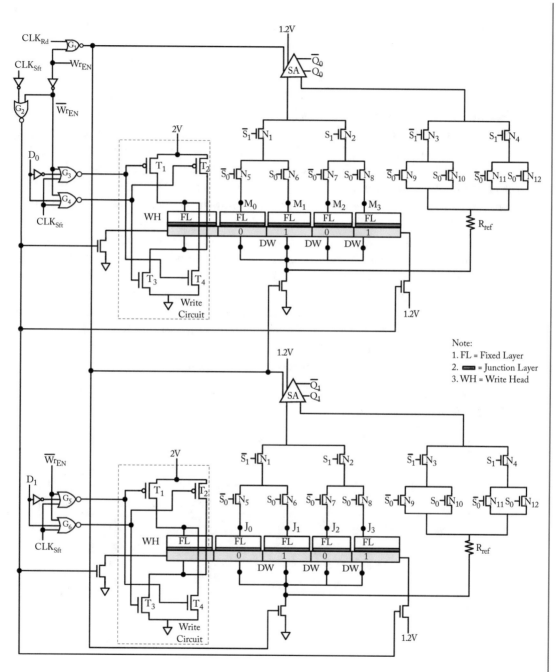

Figure 2.24: RM-based 2-input reconfigurable encoder/decoder.

at the ends of the Racetrack and the current pulses are injected into the Racetrack. As a result, DWs in the Racetrack are shifted and the bit in the write-head is pushed into the Racetrack along with the other bits. In this way, the store operations are performed and the coding scheme can be reconfigured.

When $Wr_{EN} = 0$, output of gates $G2$–$G6$ are 0 and so, all the write and DW-shift circuits are in off state. Output of $G1$ is HIGH, driving the SAs to evaluation phase. In this phase, the selected MTJ on the RM is read and its state is produced as output. A binary tree-connection of nMOS transistors is used at the left branch of the SAs to select the MTJ in RM corresponding to the input to the encoder. To balance load, an identical select tree has been added at the right branch of the SAs. For example, if the input to the encoder is 00, transistors $N1$ and $N5$ are activated in each of the SAs. As a result, the MTJs $M0$ and $J0$ are selected and their states are produced as output bits of the encoder. Therefore, only the read operations are performed when $Wr_{EN} = 0$. This design can also be used as decoder.

The above encoder/decoder implementation is suitable for realizing encoding schemes whose truth table does not remain fixed, but vary over the period. The reconfiguration feature of the encoder, however, is not required in schemes which have a fixed truth table. The reconfiguration part of the encoder consists of circuits for writing into the write-head of RTs and for DW motion. They can incur high dynamic and leakage power, especially due to the larger width of transistors in write circuits. Figure 2.25 shows a modified encoder that is suitable for fixed encoding schemes. The output bits of this encoder are initially stored in the RTs. During the encoding operation, the bits corresponding to the inputs are read using an SA as shown.

Table 2.6: Simulation parameters

Parameters	Description	Values
V_{write}	Writing voltage	2 V [Harms et al., 2010]
W_{write}	Width of transistors in writing circuits	0.4 μm
V_{read}	Reading voltage	1.2 V [Harms et al., 2010]
W_{read}	Width of transistors in read circuits	0.135 μm
I_{shift}	Amplitude of shifting pulses	98.6 μA
a	Length of MTJs	40 nm
b	Width of MTJs	40 nm

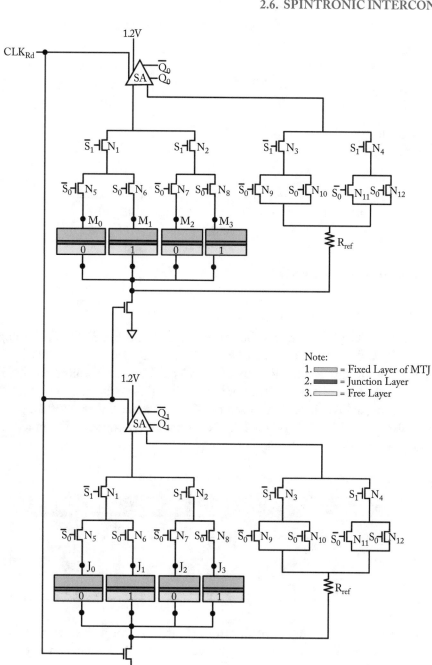

Figure 2.25: RM-based 2-input encoder/decoder for fixed encoding scheme.

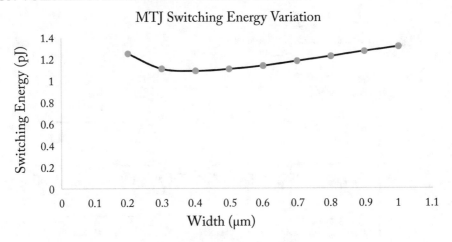

Figure 2.26: MTJ switching energy variation.

2.6.3   PERFORMANCE EVALUATION

ST Microelectronics 65 nm Process Design Kit (PDK) and the SPINLIB Verilog-A compact models for Perpendicular Magnetic Anisotropy RM [Zhang et al., 2012b] and MTJ [Zhang et al., 2012a] were used to simulate the RM-based encoder in Cadence Virtuoso Analog Design Environment. The models are based on current induced STT for switching of CoFeB/MgO/CoFeB MTJ and shifting of DWs in CoFeB racetrack. Values of some of the important parameters of the circuit are listed in Table 2.7 Synopsis DC tool and Global Foundries 65 nm technology library were used to synthesize CMOS-only encoders.

Table 2.7: Leakage power of reconfigurable encoder/decoder

Input Size	RM-based (μW)	CMOS only (μW)
1 – *bit*	0.987	0.057
2 – *bit*	1.930	0.155
4 – *bit*	3.81	0.555
6 – *bit*	5.735	2.5

We simulated the write circuit shown in Figure 2.26 for various width of transistors and plotted the corresponding switching energy. The variation of switching energy is shown in Figure 2.26. As can be seen, the transistor-width corresponding to minimum switching energy is 0.4 um. We utilized transistors of this width to implement the write circuits in the encoder module. The switching delay corresponding to 0.4 um width is 1.54 ns. We utilized DW-shift pulses of amplitude 98.644 μA.

Table 2.8: Dynamic power of reconfigurable encoder/decoder

Input Size	RM-based (μW)	CMOS only (μW)
1 – bit	412	180.3
2 – bit	829.25	427.6
4 – bit	1700	1416
6 – bit	2492.32	5620

Table 2.9: Operating frequency of reconfigurable encoder/decoder

Input Size	RM-based (GHz)	CMOS only (GHz)
1 – bit	2.5	1
2 – bit	2.5	1
4 – bit	2.5	1
6 – bit	0.625	1

We implemented RM-based and CMOS-only encoders/decoders of various input sizes and compared the values of their key performance parameters. Tables 2.9–2.12 outline these parameters for bit-widths. The leakage power of RM-based reconfigurable designs is more than that of CMOS only designs for all bit-widths. This is because of the write circuits which use transistors of larger width. The smaller designs consume more dynamic power than their CMOS counterparts. The 6-bit reconfigurable design performs better than the CMOS-only implementation in terms of dynamic power. However, the operating speed of the 6-bit design is considerably less and will be bottleneck to the high speed operation of the CMOS circuitry at the input of the encoder. The slow design is due to the larger tree of select transistors in its sensing amplifiers. Larger tree leads to larger capacitance, which requires more time to be charged or discharged during the pre-charge and evaluation phases. The operating speed of smaller encoder/decoder is however faster than the CMOS counterparts.

The absence of the power-consuming write circuits significantly improves the performance of the non-reconfigurable encoders. In fact, the performance of the RM-based non-reconfigurable designs is better than the CMOS-only counterparts. As shown in Tables 2.7 and 2.8, the RM-based designs show 1.1X—1.8X and 5.4X—17.34X improvement in leakage and dynamic power over their CMOS counterparts. The operating frequencies of these designs are listed in Table 2.9.

Table 2.10: Leakage power of non-reconfigurable encoder/decoder

Input Size	RM-based (μW)	CMOS only (μW)
1 − *bit*	0.025	0.045
2 − *bit*	0.047	0.088
4 − *bit*	0.105	0.183
6 − *bit*	0.183	0.276
8 − *bit*	0.334	0.368

Table 2.11: Dynamic power of non-reconfigurable encoder/decoder

Input Size	RM-based (μW)	CMOS only (μW)
1 − *bit*	8.846	153.4
2 − *bit*	25.67	310.35
4 − *bit*	105.3	609
6 − *bit*	165.7	906.81
8 − *bit*	218.8	1220

Table 2.12: Operating frequency of non-reconfigurable encoder/decoder

Input Size	RM-based (GHz)	CMOS only (GHz)
1 − *bit*	2.5	1
2 − *bit*	2.5	1
4 − *bit*	2.5	1
6 − *bit*	2	1
8 − *bit*	2	1

2.7 REFERENCES

Baibich, M. N., Broto, J., Fert, A., Van Dau, F. N., Petroff, F., Etienne, P., Creuzet, G., Friederich, A., and Chazelas, J. (1988) Giant magnetoresistance of (001) fe/(001) cr magnetic superlattices. *Physical Review Letters*, 61(21), p. 2472. DOI: 10.1103/physrevlett.61.2472. 37

Benini, L., De Micheli, G., Macii, E., Sciuto, D., and Silvano, C. (1997) Asymptotic zero-transition activity encoding for address busses in low-power microprocessor-based systems. In *VLSI, 1997. Proc. Seventh Great Lakes Symposium on, IEEE*, pp. 77–82. DOI: 10.1109/glsv.1997.580414. 68

Benini, L., Macii, A., Poncino, M., and Scarsi, R. (2000) Architectures and synthesis algorithms for power-efficient bus interfaces. *IEEE Transactions on Computer-Aided Design of Integrated Circuits and Systems*, 19(9), pp. 969–980. DOI: 10.1109/43.863637. 68

Bernevig, B. A., Hughes, T. L., and Zhang, S. C. (2006) Quantum spin hall effect and topological phase transition in hgte quantum wells. *Science*, 314(5806), pp. 1757–1761. DOI: 10.1126/science.1133734. 40

Brinkman, W., Dynes, R., and Rowell, J. (1970) Tunneling conductance of asymmetrical barriers. *Journal of Applied Physics*, 41(5), pp. 1915–1921. DOI: 10.1063/1.1659141. 36, 37

Brüne. C., Liu, C., Novik, E., Hankiewicz, E., Buhmann, H., Chen, Y., Qi, X., Shen, Z., Zhang, S., and Molenkamp, L. (2011) Quantum hall effect from the topological surface states of strained bulk hgte. *Physical Review Letters*, 106(12), pp. 126, 803. DOI: 10.1103/physrevlett.106.126803. 41

Burd, T. and Peters, B. (1994) A power analysis of a microprocessor: A study of an implementation of the mips r3000 architecture. *University California at Berkeley, Technical Report.* 67

Chen, Y., Analytis, J., Chu, J. H., Liu, Z., Mo, S. K., Qi, X. L., Zhang, H., Lu, D., Dai, X., Fang, Z., et al. (2009) Experimental realization of a three-dimensional topological insulator, bi2te3. *Science*, 325(5937), pp. 178–181. DOI: 10.1126/science.1173034. 40

Chen, Y., Li, H., Wang, X., Zhu, W., Xu, W., and Zhang, T. (2010) A nondestructive self-reference scheme for spin-transfer torque random access memory (stt-ram). In *Design, Automation and Test in Europe Conference and Exhibition (DATE), IEEE*, pp. 148–153. DOI: 10.1109/date.2010.5457219. 59, 60

Cheng, P., Song, C., Zhang, T., Zhang, Y., Wang, Y., Jia, J. F., Wang, J., Wang, Y., Zhu, B. F., Chen, X., et al. (2010) Landau quantization of topological surface states in bi_ {2} se_ {3}. *Physical Review Letters*, 105(7), pp. 076, 801. DOI: 10.1103/physrevlett.105.076801. 41

Chiba, D., Yamada, G., Koyama, T., Ueda, K., Tanigawa, H., Fukami, S., Suzuki, T., Ohshima, N., Ishiwata, N., Nakatani, Y., et al. (2010) Control of multiple magnetic domain walls by current in a co/ni nano-wire. *Applied Physics Express* 3(7), p. 3004. DOI: 10.1143/apex.3.073004. 54

Engel, B., Akerman, J., Butcher, B., Dave, R., DeHerrera, M., Durlam, M., Grynkewich, G., Janesky, J., Pietambaram, S., Rizzo, N., et al (2005) A 4-mb toggle mram based on a novel bit and switching method. *Magnetics, IEEE Transactions on*, 41(1), pp. 132–136. DOI: 10.1109/tmag.2004.840847. 45

Fei, W., Yu, H., Zhang, W., and Yeo, K. S. (2012) Design exploration of hybrid cmos and memristor circuit by new modified nodal analysis. *IEEE Transactions on Very Large Scale Integration Systems*, 20(6), pp. 1012–1025. DOI: 10.1109/tvlsi.2011.2136443. 31

Fu, L., Kane, C. L., and Mele, E. J. (2007) Topological insulators in three dimensions. *Physical Review Letters*, 98(10), pp. 106, 803. DOI: 10.1103/physrevlett.98.106803. 40

Fujita, T., Jalil, M. B. A., and Tan, S. G. (2011) Topological insulator cell for memory and magnetic sensor applications. *Applied Physics Express* 4(9), pp. 094, 201. http://apex.jsap.jp/link?APEX/4/094201/ DOI: 10.1143/apex.4.094201. 41, 43

Fukami, S., Suzuki, T., Nagahara, K., Ohshima, N., Ozaki, Y., Saito, S., Nebashi, R., Sakimura, N., Honjo, H., Mori, K., et al. (2009) Low-current perpendicular domain wall motion cell for scalable high-speed mram. In *Symposium on VLSI Technology*. 54

Grünberg, P., Schreiber, R., Pang, Y., Brodsky, M., and Sowers, H. (1986) Layered magnetic structures: Evidence for antiferromagnetic coupling of fe layers across cr interlayers. *Physical Review Letters*, 57(19), p. 2442. DOI: 10.1103/physrevlett.57.2442. 36

Harms, J. D., Ebrahimi, F., Yao, X., and Wang, J. P. (2010) Spice macromodel of spin-torque-transfer-operated magnetic tunnel junctions. *Electron Devices, IEEE Transactions on*, 57(6), pp. 1425–1430. DOI: 10.1109/ted.2010.2047073. 38

Ho, C. W., Ruehli, A., and Brennan, P. (1975) The modified nodal approach to network analysis. *Circuits and Systems, IEEE Transactions on*, 22(6), pp. 504–509. DOI: 10.1109/tcs.1975.1084079. 31

Hosomi, M., Yamagishi, H., Yamamoto, T., Bessho, K., Higo, Y., Yamane, K., Yamada, H., Shoji, M., Hachino, H., Fukumoto, C., et al. (2005) A novel nonvolatile memory with spin torque transfer magnetization switching: Spin-ram. In *Electron Devices Meeting. IEDM Technical Digest. IEEE International*, pp. 459–462. DOI: 10.1109/iedm.2005.1609379. 39

Hsieh, D., Qian, D., Wray, L., Xia, Y., Hor, Y. S., Cava, R., and Hasan, M. Z. (2008) A topological dirac insulator in a quantum spin hall phase. *Nature*, 452(7190), pp. 970–974. DOI: 10.1038/nature06843. 40

Hsieh, D., Xia, Y., Qian, D., Wray, L., Dil, J., Meier, F., Osterwalder, J., Patthey, L., Checkelsky, J., Ong, N., et al. (2009) A tunable topological insulator in the spin helical dirac transport regime. *Nature*, 460(7259), pp. 1101–1105. DOI: 10.1038/nature08234. 40

ITRS (2010) International technology roadmap of semiconductor. http://www.itrs.net 50

Jeong, G., Cho, W., Ahn, S., Jeong, H., Koh, G., Hwang, Y., and Kim, K. (2003) A 0.24-μm 2.0-v 1t1mtj 16-kb nonvolatile magnetoresistance ram with self-reference sensing scheme. *Solid-State Circuits, IEEE Journal of*, 38(11), pp. 1906–1910. DOI: 10.1109/jssc.2003.818145. 57, 59, 60

Kawahara, T., Takemura, R., Miura, K., Hayakawa, J., Ikeda, S., Lee, Y., Sasaki, R., Goto, Y., Ito, K., Meguro, I., et al. (2007) 2mb spin-transfer torque ram (spram) with bit-by-bit bidirectional current write and parallelizing-direction current read. In *Solid-State Circuits Conference, ISSCC. Digest of Technical Papers. IEEE International*, pp. 480–617. DOI: 10.1109/isscc.2007.373503. 39

Koch, R., Deak, J., Abraham, D., Trouilloud, P., Altman, R., Lu, Y., Gallagher, W., Scheuerlein, R., Roche, K., Parkin, S. (1998) Magnetization reversal in micron-sized magnetic thin films. *Physical Review Letters*, 81(20), p. 4512. DOI: 10.1103/physrevlett.81.4512. 46, 47, 48

Liu, D. and Svensson, C. (1994) Power consumption estimation in cmos vlsi chips. *IEEE Journal of Solid-State Circuits*, 29(6), pp. 663–670. DOI: 10.1109/4.293111. 67

Moore, J. E. (2010) The birth of topological insulators. *Nature*, 464(7286), pp. 194–198. DOI: 10.1038/nature08916. 40

Musoll, E., Lang, T., and Cortadella, J. (1998) Working-zone encoding for reducing the energy in microprocessor address buses. *IEEE Transactions on Very Large Scale Integration (VLSI) Systems*, 6(4), pp. 568–572. DOI: 10.1109/92.736129. 67, 68

Nagel, L. W. and Pederson, D. O. (1973) SPICE: Simulation program with integrated circuit emphasis. *Electronics Research Laboratory*, College of Engineering, University of California. 31

Nenzi, P. and Holger, V. (2010) Ngspice Users Manual. http://www.itrs.net 46

Parkin, S. S., Hayashi, M., and Thomas, L. (2008) Magnetic domain wall racetrack memory. *Science*, 320(5873), pp. 190–194. DOI: 10.1126/science.1145799. 50

Qi, X. L., Wu, Y. S., and Zhang, S. C. (2006) Topological quantization of the spin hall effect in two-dimensional paramagnetic semiconductors. *Physical Review B*, 74(8), pp. 085, 308. DOI: 10.1103/physrevb.74.085308. 43

Ramprasad, S., Shanbhag, N. R., and Hajj, I. N. (1999) A coding framework for low-power address and data busses. *IEEE Transactions on Very Large Scale Integration (VLSI) Systems*, 7(2), pp. 212–221. DOI: 10.1109/92.766748. 68

Rowlands, G., Rahman, T., Katine, J., Langer, J., Lyle, A., Zhao, H., Alzate, J., Kovalev, A., Tserkovnyak, Y., Zeng, Z., et al, (2011) Deep subnanosecond spin torque switching in magnetic tunnel junctions with combined in-plane and perpendicular polarizers. *Applied Physics Letters*, 98(10), pp. 102, 509. DOI: 10.1063/1.3565162. 61, 62

Sakurai, T. (2000) Design challenges for 0.1 um and beyond: embedded tutorial. In *Proc. of the 2000 Asia and South Pacific Design Automation Conference, ACM*, pp. 553–558. DOI: 10.1145/368434.368787. 67

Shang, Y., Fei, W., and Yu, H. (2012) Analysis and modeling of internal state variables for dynamic effects of nonvolatile memory devices. *IEEE Transactions on Circuits and Systems I Regular Papers*, 59(9), pp. 1906–1918. DOI: 10.1109/tcsi.2011.2180441. 31

Sotiriadis, P. P. and Chandrakasan, A. P. (2002) A bus energy model for deep submicron technology. *IEEE Transactions on Very Large Scale Integration (VLSI) Systems*, 10(3), pp. 341–350. DOI: 10.1109/tvlsi.2002.1043337. 67

Stan, M. R. and Burleson, W. P. (1995) Bus-invert coding for low-power i/o. *IEEE Transactions on Very Large Scale Integration (VLSI) Systems*, 3(1), pp. 49–58. DOI: 10.1109/92.365453. 68

Stan, M. R. and Burleson, W. P. (1997) Low-power encodings for global communication in cmos vlsi. *IEEE Transactions on Very Large Scale Integration (VLSI) Systems*, 5(4), pp. 444–455. DOI: 10.1109/92.645071. 68

Thomas, L., Yang, S. H., Ryu, K. S., Hughes, B., Rettner, C., Wang, D. S., Tsai, C. H., Shen, K. H., and Parkin, S. S. (2011) Racetrack memory: a high-performance, low-cost, non-volatile memory based on magnetic domain walls. In *Electron Devices Meeting (IEDM), IEEE International*, pp. 24–2. DOI: 10.1109/iedm.2011.6131603. 50

Thoziyoor, S., Muralimanohar, N., Ahn, J. H., and Jouppi, N. P. (2008) Cacti 5.1. HP Laboratories, April 2. 65

Trinh, H. P., Zhao, W., Klein, J. O., Zhang, Y., Ravelsona, D., and Chappert, C. (2012) Domain wall motion based magnetic adder. *Electronics Letters*, 48(17), pp. 1049–1051. DOI: 10.1049/el.2012.1577. 63

Venkatesan, R., Kozhikkottu, V., Augustine, C., Raychowdhury, A., Roy, K., and Raghunathan, A. (2012) Tapecache: a high density, energy efficient cache based on domain wall memory. In *Proc. of the 2012 ACM/IEEE International Symposium on Low Power Electronics and Design, ACM*, pp. 185–190. DOI: 10.1145/2333660.2333707. 50

Wang, K. L., Zhao, Z., and Khitun, A. (2008) Spintronics for nanoelectronics and nanosystems. *Thin Solid Films*, 517(1), pp. 184–190. DOI: 10.1016/j.tsf.2008.08.145. 40

Wang, X., Zhu, W., Siegert, M., and Dimitrov, D. (2009) Spin torque induced magnetization switching variations. *Magnetics, IEEE Transactions on*, 45(4), pp. 2038–2041. DOI: 10.1109/tmag.2009.2015376. 36

Wang, Y., Yu, H., Sylvester, D., and Kong, P. (2012) Energy efficient in-memory AES encryption based on nonvolatile domain wall nanowire, *Design, Automation and Test in Europe Conference and Exhibition (DATE)*, IEEE, pp. 1–4. DOI: 10.7873/date.2014.196.

Wolf, S., Awschalom, D., Buhrman, R., Daughton, J., Von Molnar, S., Roukes, M., Chtchelka-nova, A. Y., and Treger, D. (2001) Spintronics: A spin-based electronics vision for the future. *Science*, 294(5546), pp. 1488–1495. DOI: 10.1126/science.1065389. 40

Xia, Y., Qian, D., Hsieh, D., Wray, L., Pal, A., Lin, H., Bansil, A., Grauer, D., Hor, Y., Cava, R., et al. (2009) Observation of a large-gap topological-insulator class with a single dirac cone on the surface. *Nature Physics*, 5(6), pp. 398–402. DOI: 10.1038/nphys1274. 40

Zhang, Y., Zhao, W., Lakys, Y., Klein, J. O., Kim, J. V., Ravelosona, D., and Chappert, C. (2012a) Compact modeling of perpendicular-anisotropy cofeb/mgo magnetic tunnel junctions. *IEEE Transactions on Electron Devices*, 59(3), pp. 819–826. DOI: 10.1109/ted.2011.2178416. 72

Zhang, Y., Zhao, W., Ravelosona, D., Klein, J. O., Kim, J., and Chappert, C. (2012b) Perpendicular-magnetic-anisotropy cofeb racetrack memory. *Journal of Applied Physics*, 111(9), pp. 093, 925. DOI: 10.1063/1.4716460. 72

Zhao, H., Lyle, A., Zhang, Y., Amiri, P., Rowlands, G., Zeng, Z., Katine, J., Jiang, H., Galatsis, K., Wang, K., et al. (2011) Low writing energy and sub nanosecond spin torque transfer switching of in-plane magnetic tunnel junction for spin torque transfer random access memory. *Journal of Applied Physics*, 109(7), p. 07C720. DOI: 10.1063/1.3556784. 61, 62

Zhao, H., Glass, B., Amiri, P. K., Lyle, A., Zhang, Y., Chen, Y. J., Rowlands, G., Upadhyaya, P., Zeng, Z., Katine, J., et al. (2012) Sub-200 ps spin transfer torque switching in in-plane magnetic tunnel junctions with interface perpendicular anisotropy. *Journal of Physics D: Applied Physics*, 45(2), pp. 025, 001. DOI: 10.1088/0022-3727/45/2/025001. 61, 62

Zhao, W., Chappert, C., Javerliac, V., and Noziere, J. P. (2009) High speed, high stability and low power sensing amplifier for mtj/cmos hybrid logic circuits. *IEEE Transactions on Magnetics*, 45(10), pp. 3784–3787. DOI: 10.1109/tmag.2009.2024325. 68

Zyuzin, A. and Burkov, A. (2011) Thin topological insulator film in a perpendicular magnetic field. *Physical Review B*, 83(19), pp. 195, 413. DOI: 10.1103/physrevb.83.195413. 41

CHAPTER 3

In-memory Data Encryption

Abstract   Big-data storage poses significant challenges to anonymization of sensitive information against data sniffing. Not only will the encryption bandwidth be limited by the I/O traffic, the transfer of data between processor and memory will also expose the input-output mapping of intermediate computations on I/O channels that are susceptible to semi-invasive and non-invasive attacks. Limited by the simplistic cell-level logic, existing logic-in-memory computing architectures are incapable of performing the complete encryption process within the memory at reasonable throughput and energy efficiency. The Advanced Encryption Standard (AES), also known as Rijndael, is a specification for the encryption of electronic data. However, it takes several hours to encrypt/decrypt one Terabyte of data in 128-bit AES on a general purpose computer. While cheap distributed storage is the driving factor behind big data, the cost for keeping its data confidentiality can be intimidating. Limited by footprint and energy dissipation, data encryption is less affordable on resource constrained devices. In this regard, area-efficient, low-power, and high throughput in-memory encryption engine is of great significance not only for privacy in big data processing, but also on resource-constrained environments. In addition, another encryption approach Simon cipher also shows benefits in data encryption. In this chapter, both AES and Simon cipher are implemented by the domain wall nanowire based in-memory architecture.

3.1   IN-MEMORY ADVANCED ENCRYPTION STANDARD

3.1.1   FUNDAMENTAL OF AES

The Advanced Encryption Standard (AES) algorithm operates on plaintext block of 128 bits. The 16-byte input data is internally organized into an array of four rows by four columns, called the *state matrix* (M_s). The input data is encrypted by applying a sequence of transformations to the state matrix, as detailed by the pseudo code in Algorithm 3.1. The flow chart in Figure 3.1 shows the dominant resources used for the hardware implementation of each transformation module, where the gate utilization data is obtained by synthesizing the AES Verilog code from Usselmann [2002]. The operations performed and the resources consumed for each module are described as follows.

- SubBytes: Each byte $S_{i,j}$ of the state matrix M_s will be independently updated by a non-linear transformation f in this module. The mapping f is performed by a substitution-box (S-box), which takes one byte of input from M_s and transforms it into another byte at the same position. The SubBytes module accounts for half of the total gates in AES, with regis-

---

Algorithm 3.1 Advanced encryption standard

Require: 128-bit plaintext
1: Organize input data as 4×4 state matrix M_s, each entry $S_{i,j}$ has one byte
2: **for** $r = 1 : N_r$ **do**
3:   **for all** $S_{i,j} \in M_s$ **do**
4:     SubBytes transformation: $S'_{i,j} \leftarrow f(S_{i,j})$
5:   **end for**
6:   **for all** $S_{n,j} \in M_s$ in n_{th} row, $n \in \{1, 2, 3, 4\}$ **do**
7:     $k \leftarrow$ (j-n+1) mod 4
8:     ShiftRows transformation: $S'_{n,j} \leftarrow S_{n,k}$, i.e., each row is left shifted circularly by $n - 1$ bytes
9:   **end for**
10:   **if** $i \neq N_r$ **then**
11:     MixColumns transformation: $M_s \leftarrow M_s \times M_{mc}$
12:   **end if**
13:   **for all** $S_{i,j} \in M_s$ **do**
14:     AddRoundKey transformation: $S'_{i,j} \leftarrow S_{i,j} \oplus K_{i,j}$
15:   **end for**
16:   $M_s \leftarrow M'_s$
17: **end for**
Ensure: 128-bit ciphertext

---

ters used as fixed storage elements of the look-up table (LUT). Each S-box is pre-configured with an 8-bit word in each memory location addressable by an 8-bit input. Hence, the LUT size is $2^8 \cdot 8 = 2048$ bits. The percentage of hardware resources utilized by this module may vary depending on how the S-box is implemented. If the S-box is implemented by combinational logic circuit, XOR gates become the dominant resources, which account for more than 70% of gate utilization for the AES implementation, as reported in Abid et al. [2009].

- ShiftRows: The n_{th} row of M_s will be cyclically shifted to the left by n bytes. As shown by arrows on the entries of M_s in Figure 3.1, the top row is not shifted; the second row is shifted by one byte position; the third row by two; and the fourth row by three. In ASIC design, the ShiftRows transformation can be performed in-place by storing the content of M_s in shift registers, hence no additional logic gate is incurred.

- MixColumns: Each column of M_s is multiplied by a constant matrix M_{mc} consisting of three integer values 1, 2, and 3, as shown in Figure 3.1. The multiplication of a variable byte by an integer 1, 2, or 3 of M_{mc} results in an unchanged byte, a 1-bit left shifted byte or the XOR of the byte with a 1-bit left shifted version of itself, respectively. Hence, this invertible linear transformation replaces all four bytes in a column of M_s such that each byte is mixed with all four bytes in the column. This module contains only XOR gates and it accounts for nearly half of the total number of gates.

- AddRoundKey: The 16-byte round keys are organized in a similar 4×4 array M_k as the state matrix. Each entry of M_k is denoted as $K_{i,j}$. In this operation, each byte $S_{i,j}$ of M_s will be replaced by the result of a bitwise-XOR operation with a byte $K_{i,j}$ in the same row and column of the round key matrix M_k. Therefore, the AddRoundKey module is again built by merely XOR gates, which accounts for 3.3% of the total gates.

In conclusion, the basic operations involved in AES are XOR, shift, and table look up although the percentage utilization of the logic gates used for each module may vary from design to design.

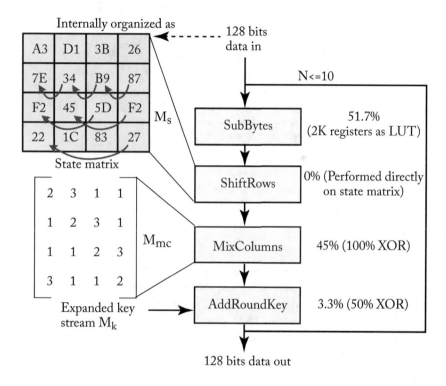

Figure 3.1: Flow chart of AES algorithm with gate utilization analysis.

3.1.2 DOMAIN WALL NANOWIRE-BASED AES COMPUTING

Data Organization of State Matrix

Because in-memory encryption is performed directly on the data cells, the data needs to be organized in certain fashion to facilitate the execution of AES algorithm [Wang et al., 2016]. As discussed in Chapter 2, DW nanowires can only support serial access, i.e., one bit of information can be accessed from a DW nanowire at any one time. In order to access multiple bits within one cycle, the data needs to be distributed into separate nanowires so that they can be operated concurrently. In AES algorithm, the basic processing unit is one byte of the state matrix M_s. Therefore, M_s is split into eight 4×4 binary arrays $A_k \ \forall k \in [0, 7]$, as illustrated in Figure 3.2.

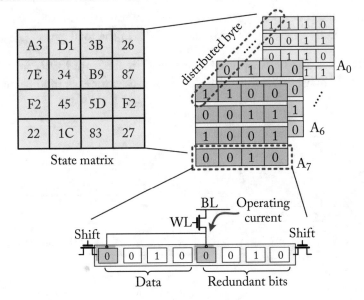

Figure 3.2: Distributed data organization of state matrix for in-memory operation by domain wall nanowire devices.

The binary bit in the i-th row and j-th column of each array A_k corresponds to the k-th bit of the state byte $S_{i,j}$ of M_s. By distributing the Boolean operations on state bytes $S_{i,j}$ into eight binary arrays, the byte access requirement of AES algorithm is satisfied.

   To exploit the shift property of DW nanowire for the ShiftRows transformation of AES, each row of the binary array needs to be stored in one DW nanowire. Thus, four nanowires are needed for each array A_k, $k \in [0, 7]$. In each nanowire, some redundant bits are appended to the four data bits to reduce the shift current required for the circular left shift operation. The rationale will be elaborated later in the implementation of ShiftRows transformation.

SubBytes

Instead of using SRAM LUT to implement the S-box, we propose implementing the S-box by non-volatile DW nanowire devices, abbreviated as DW-LUT, as shown in Figure 3.3. The 8-bit results are split into separated nanowires to speed up this nonlinear transformation. Eight SAs are required for each DW-LUT in order to output the result parallelly. As both the in-memory state matrix and S-box are made up of non-volatile DW nanowires, leakage power is reduced significantly. Further area and power savings are achieved by sharing the decoders and SAs for the memory and DW-LUT. The number of cycles consumed in the *SubBytes* stage, t_{SB}, can be derived as follows:

$$t_{SB} = (t_{read} + t_{LUT} + t_{write}) \cdot \frac{16}{N_{LUT}}, \tag{3.1}$$

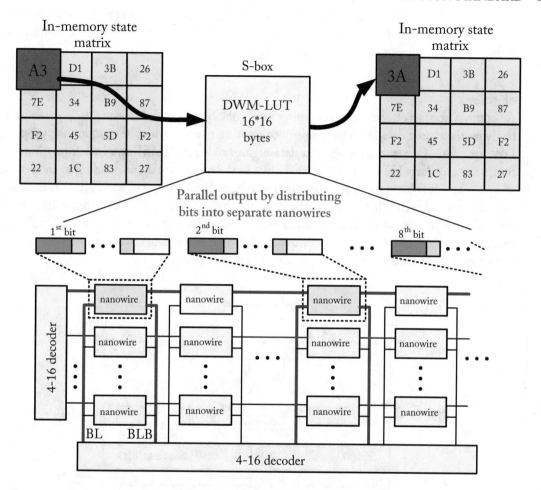

Figure 3.3: SubBytes stage with S-box function realized by domain wall, memory-based look-up table.

where $N_{LUT} \in \{1, 2, 4\}$, t_{read} and t_{write} are the read and write latency (in number of clock cycles) of state matrix M_s, and t_{LUT} is the latency of look-up table in number of cycles.

As each DW-LUT handles one $S_{i,j}$ transformation at a time, the SubBytes transformations can be parallelized by using more DW-LUT. The maximum value of N_{LUT} is limited by the number of bytes that can be accessed in the state matrix M_s. As each A_k can have at most four SAs, one to each nanowire for each row of A_k, at most four bytes can be accessed simultaneously. The SAs can be shared among different nanowires to save area if full parallelism is not required.

The composite fields for S-box and inverse S-box operations can be represented using normal basis, polynomial basis, or mixed basis [Nogami et al., 2010]. In the proposed DW nanowire architecture, DW-LUT instead of combinational logic is used to store the transformed data of SubBytes and InvSubBytes, hence the hardware complexity of their implementation is not affected by using different bases for the mapping to composite field S-boxes.

ShiftRows

As opposed to the conventional computing flow, the ShiftRows transformation is performed directly on the stored data in an in-memory computing manner with the unique shift property of DW nanowire. Due to the distributed data organization, the ShiftRows transformation is performed by circularly left shifting the binary bits in the nanowires corresponding to the i-th rows of all A_k by i bits $\forall i \in \{0, 1, 2, 3\}$. In other words, the second row of each bit matrix A_k needs to be left shifted cyclically by one bit, the third row by two bits, and the fourth row by three bits, while the top row remains unchanged. To avoid writing back the most significant bit (MSB) to the least significant bit (LSB) for each circular left shift operation, redundant bits are padded to form a virtual circle on the nanowire, as illustrated in Figure 3.4.

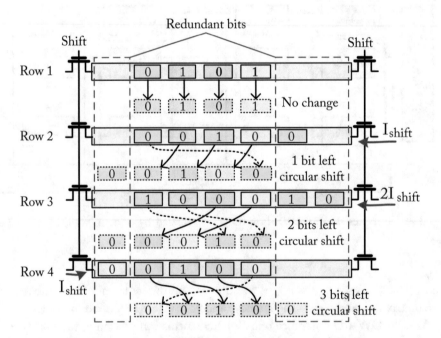

Figure 3.4: ShiftRows transformation by domain wall nanowire shift operations.

The number of redundant bits (domains) to be added in each nanowire is determined by the number of bits to be circularly left shifted for each row. As the shift operation for each nanowire can be performed concurrently, the number of cycles consumed in the *ShiftRows* stage, t_{SR}, can be determined by the longest row shift delay, i.e.,

$$t_{SR} = \max(t_{rs_0}, t_{rs_1}, t_{rs_2}, t_{rs_3}), \tag{3.2}$$

where t_{rs_i} denotes the row shift delay of the i-th row, $i \in \{0, 1, 2, 3\}$.

In order to complete all shifts in one cycle, shift current of different amplitudes is applied to each row according to the linear current-velocity relationship of shift operation [Augustine et al., 2011]. This implies that the amplitudes of the shift currents applied to the nanowires for the third and fourth rows are twice and three times more than that applied to the nanowire for the second row. To reduce the shift current, consider the following equivalent circular left and right shift operations of a four-bit operand: $LS(1) \overset{\text{def}}{=\!=} RS(3), LS(2) \overset{\text{def}}{=\!=} RS(2), LS(3) \overset{\text{def}}{=\!=} RS(1)$, where $LS(i)$ and $RS(i)$ denote the left and right shift operations with the number of bits to be shifted indicated by the integer argument i. This means that instead of shifting three bits leftward for the last row, only one bit right shifting is required, which reduces the redundant data from three bits to one bit and the amplitude of the applied shift current to only one third of what is originally required. In Figure 3.4, the bits in the same color are synchronized bits. To ensure that the correct results will always be obtained by the circular shift, the redundant bits must also be updated coherently as the state matrix changes.

AddRoundKey

In the AddRoundKey stage, each byte in the state array will be bitwise XORed with the corresponding key byte. As the key logic operation in this stage is XOR, we propose a nanowire-based XOR logic (DW-XOR) for leakage free computing. As described in Chapter 2, the GMR-effect can be considered as an XOR operation of the magnetization directions of two thin magnetic layers, where the output is determined by the high or low resistance value measured by the MTJ. In a GMR-based MTJ structure, however, only one operand of the XOR logic can be variable as the magnetization of the fixed layer is constant. This problem can be overcome by constructing a new read-only-port, where two free layers and one insulator layer are stacked.

The two free layers each have the size of a magnetization domain and are obtained from two different nanowires. Thus, the two operands, representing the magnetization direction in each free layer, can both be variables. Their values can be assigned through the MTJs of their own nanowire, and then shifted to the operating port to carry out the XOR operation.

The bitwise-XOR logic for the AddRoundKey operation implemented by two DW nanowires is shown in Figure 3.5. Let A and B be the 1-bit operands from the state byte and key byte, respectively. Eight identical DW-XORs are used for the bitwise XOR operation between the state byte and the key byte. The following steps are involved in the execution of the operation *state* $\oplus$ *key*.

- The binary operands, A and B, are loaded into two nanowires by enabling WR_1 and WR_2, respectively.

- *A* and *B* are shifted from their access-ports to the read-only-port by enabling SHF_1 and SHF_2, respectively.

- By enabling *RD*, the bitwise-XOR result can be obtained from the read-only-port through the GMR-effect.

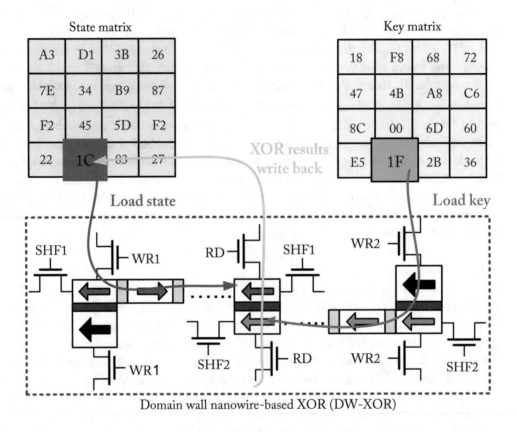

Figure 3.5: AddRoundKey stage with XOR logic realized by domain wall nanowire.

The latency t_{ARK} in terms of the number of cycles consumed in the *AddRoundKey* stage can be calculated by:

$$t_{ARK} = (t_{read} + t_{xor} + t_{write}) \cdot \frac{128}{N_{xor}}, \tag{3.3}$$

where $N_{xor} \in \{1, 2, 4, 8, 16, 32\}$.

Similar to the *SubBytes* transformation, using more DW-XOR units and SAs will increase the parallelism and reduce the latency for this stage. When $N_{xor} = 4 \times 8 = 32$, the maximum parallelism is reached and one complete column of M_s is processed simultaneously in the *AddRoundKey* stage.

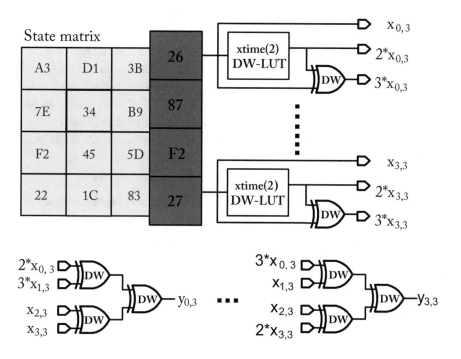

Figure 3.6: MixColumns transformation by DW-LUT and DW-XOR.

MixColumns

The MixColumns transformation can be expressed as a multiplication of the state matrix M_s by the known matrix M_{mc}, as shown in Figure 3.1. The i-th column, $i \in \{0, 1, 2, 3\}$, after this transformation becomes

$$
\begin{aligned}
S'_{0,i} &= 2 \times S_{0,i} \oplus 3 \times S_{1,i} \oplus S_{2,i} \oplus S_{3,i} \\
S'_{1,i} &= S_{0,i} \oplus 2 \times S_{1,i} \oplus 3 \times S_{2,i} \oplus S_{3,i} \\
S'_{2,i} &= S_{0,i} \oplus S_{1,i} \oplus 2 \times S_{2,i} \oplus 3 \times S_{3,i} \\
S'_{3,i} &= 3 \times S_{0,i} \oplus S_{1,i} \oplus S_{2,i} \oplus 2 \times S_{3,i},
\end{aligned}
$$

where $S_{i,j}$ and $S'_{i,j}$ denote the values of the byte in the i-th row and j-th column of the state matrix before and after the MixColumns transformation, respectively.

The MixColumns transformation can be performed merely with DW-LUT and DW-XOR, as shown in Figure 3.6. The operations involved are multiplication by two (denoted by *xtime-2*), multiplication by three (denoted by *xtime-3*) and bitwise XOR. Different from the normal integer multiplication, *xtime-2* is obtained by left shifting the operand by one bit, followed by bitwise XORing it with *0x1B*[1] if the MSB of the operand is 1 before shifting; *xtime-3*

---

[1] *0x1B* corresponds to the irreducible polynomial $x^8 + x^4 + x^3 + x + 1$ in $GF(2^8)$ with the highest degree term discarded. This is because the term with the highest degree and the carry add modulo 2 to 0.

is obtained by XORing the operand with its result of *xtime-2*. The bitwise XOR can be executed by the proposed DW-XOR. Although *xtime-2* can be implemented by in-memory shifting with additional DW-XOR, it is more efficient to be implemented by an 8-bit input, 8-bit output DW-LUT due to the associated branching operation that is dependent on the value of its MSB.

Unlike the *SubBytes* and *AddRoundKey* stages, where the transformation for each byte $S_{i,j}$ is independent of other bytes, each output byte of $S'_{i,j}$ in *MixColumns* is a weighted combination of four bytes in a column of M_s. Therefore, it cannot be further parallelized by having more DW-LUT units and DW-XOR. The number of cycles t_{MC} consumed in the *MixColumns* stage can be calculated as follows:

$$t_{MC} = (t_{read} + t_{LUT} + 3 \cdot t_{xor} + t_{write}) \cdot 4. \tag{3.4}$$

Due to the multiple-cycle operation of DW-XOR, and higher number of DW-XORs in the critical path, *MixColumns* is the most time-consuming stage among the four stages.

Algorithm 3.2 High throughput/energy efficiency domain wall nanowire-based AES mapping

Require: AES design area constraint A_m or/and power constraint P_m
1: $S_R \leftarrow \emptyset$
2: $\mathcal{R}_{SR} \leftarrow \{32 \times M_{nw}\}; \mathcal{R}_{MC} \leftarrow \{4 \times M_{lut}, 8 \times M_{4-16dec}\}$
3: **for** (N_{xor}, N_{lut}) combinations, $N_{xor} \in \{1, 2, 4...32\}$ and $N_{lut} \in \{1, 2, 4\}$ **do**
4:     $\mathcal{R}_{ARK} \leftarrow \{N_{xor} \times M_{xor}, 32 \times M_{sa}\}$
5:     $\mathcal{R}_{SB} \leftarrow \{N_{lut} \times M_{lut}, 32 \times M_{sa}, 8 \times M_{4-16dec}\}$
6:     calculate $\{t_{SB}, t_{ARK}, t_{SR}, t_{MC}\} \leftarrow f(\{\mathcal{R}_{SB}, \mathcal{R}_{ARK}, \mathcal{R}_{SR}, \mathcal{R}_{MC}\})$
7:     $t'_{cycle} \leftarrow Sum\{t_{SB}, t_{ARK}, t_{SR}, t_{MC}\}$
8:     $S'_R \leftarrow \mathcal{R}_{SB} \cup \mathcal{R}_{ARK} \cup \mathcal{R}_{SR} \cup \mathcal{R}_{MC}$
9:     calculate area $A \leftarrow g(S'_R)$, power $P \leftarrow h(S'_R)$, and throughput $T' \leftarrow j(S'_R)$
10:     **if** $A \leq A_m$ **and** $P \leq P_m$ **and** $t'_{cycle} \leq t_{cycle}$ **then**
11:         $S_R \leftarrow S'_R$
12:         $t_{cycle} \leftarrow t'_{cycle}$
13:     **end if**
14: **end for**
Ensure: DW nanowire based AES architecture resource utilization S_R

KeyExpansion

AES in Figure 3.1 assumes a 128-bit key but the input data block needs to undergo ten rounds of AddRoundKey operation. To achieve the objective of confusion, the KeyExpansion function needs to generate a different round key for every iteration of AddRoundKey operation. For a higher level of security, 192- and 256-bit keys with 12 rounds and 14 rounds of mixing of the same 128-bit input block are applied.

The KeyExpansion operation itself is composed of circular rotation, byte substitution, XOR operation with the round constant and word-level XOR [Daemen and Rijmen, 2002], which can all be decomposed into the four previously introduced basic operations. In other words, the KeyExpansion function for all three key lengths of AES can be mapped to the previously described operations of DW nanowire devices. Algorithm 3.2 provides the hardware mapping of overall AES operations to DW devices for given design constraints.

DW-AND Operation for GCM Architecture Extension

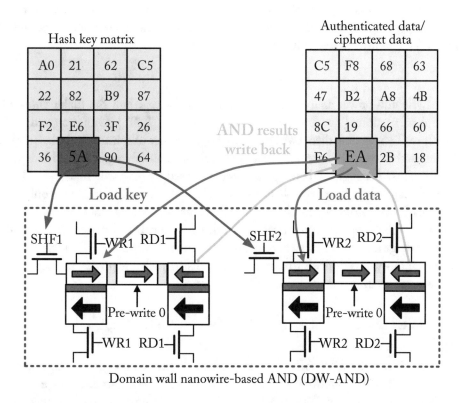

Figure 3.7: AND logic realized by domain wall nanowire for AES-GCM.

Among different modes of operation for symmetric key cryptographic block ciphers, Galois Counter Mode (GCM) has been widely adopted due to its efficiency and performance. As an efficient authenticated encryption algorithm, AES-GCM [Satoh et al., 2009] can fully exploit the parallel and pipelined processing of the original AES in Figure 3.1 to provide both data authenticity and confidentiality. Besides all the operations required in Figure 3.1, a GCM hash function is needed to generate the authentication tag. The Galois Field (GF) multiplication of GCM hash function can be obtained by ANDing the authenticated data A and the ciphertext data C with a hash key H, where A, C, and H are all 128-bit data. This 128×128 AND-array can be implemented using a DW memory based AND operation. Figure 3.7 shows the detailed implementation of a 1-bit AND operation on the DW nanowire. To perform an AND operation, we need to pre-write a 0 into a domain of the DW nanowire first, and then write a bit of A or C into the next domain. The corresponding bit in H is then applied on the shift transistor. If the hash key bit is 0, the DW nanowire will not shift and the pre-write 0 will be read out. If the data in A or C is 0, the readout data will always be 0 irrespective of whether there is any DW movement in the nanowire. Altogether, 128 pairs of 1-bit AND DW nanowire are needed. The GF sum of the partial product bits can be easily realized by the DW-XOR structure as in the AddRoundKey stage shown in Figure 3.5. Hence, the GCM operation can be fully integrated into the DW nanowire-based AES.

Security Analysis

Fault analysis and power analysis-based side-channel attacks are the major threats in CMOS AES implementations [Bertoni et al., 2003, Karri et al., 2001]. Our proposed architecture is inherently resilient against these two attacks owing to the use of spintronic devices and in-memory computing architecture. For the fault analysis attack, an external magnetic field is required to inject the fault because the logic and storage are implemented by magnetic materials. Due to the size of the nanomagnets, it is impossible to localize the applied field to influence only a single MTJ or a DW nanowire within a densely integrated magnetic memory array to precisely inject a fault into a target cell. As opposed to charge-based devices, spin-based devices are normally off, they have zero leakage and ultra-low switching power. Moreover, as all the AES logic are implemented inside the memory, it is difficult to correlate the indistinguishable power dissipation of each logic block measured in different time slots of encryption.

3.1.3   PIPELINED AES BY DOMAIN WALL NANOWIRE

Pipelined DW-AES

Figure 3.8a shows the DW-AES implementation without pipelining. There are four stages, namely *SubBytes*, *AddRoundKey*, *ShiftRows*, and *MixColumns*, and each stage has different latency. Only one of the four stages is active at one time while the other three stages are idle. The idle stages will consume leakage power even if they do not process any data. Pipelining can be introduced to exploit the idle resources and increase their utilization rate in order to achieve higher energy efficiency and throughput. For DW-AES architecture, pipelining is not as straightforward as CMOS ASIC implementation since each DW logic stage requires different numbers of cycles. Based on the latencies of 6, 2, 11, and 5 cycles for *SubBytes*, *AddRoundKey*, *ShiftRows* and *MixColumns*, respectively, the throughput of the pipeline is limited by the most time consuming ShiftRows stage. The AddRoundKey result has to wait for nine cycles before it can accept new input data from the SubBytes stage, and transfer its result to the next ShiftRows stage. Similarly, the MixColumns stage also has to wait until ShiftRows has completed its computation in order to feed the result to it.

Instead of using timers to control the data flow for pipelining, first-in-first out (FIFO) buffers adapted to different cycle delays of each stage can be designed by exploiting the tandem movement of domain walls in nanowire. Figure 3.8b shows the timing diagram of pipelined DW-AES with the use of DW-FIFOs. The DW-FIFOs between stages are illustrated in Figure 3.8c, which are essentially nanowires of different lengths. By changing the length of nanowire and shifting one domain of the nanowire per cycle, any number of cycles can be achieved. As shown

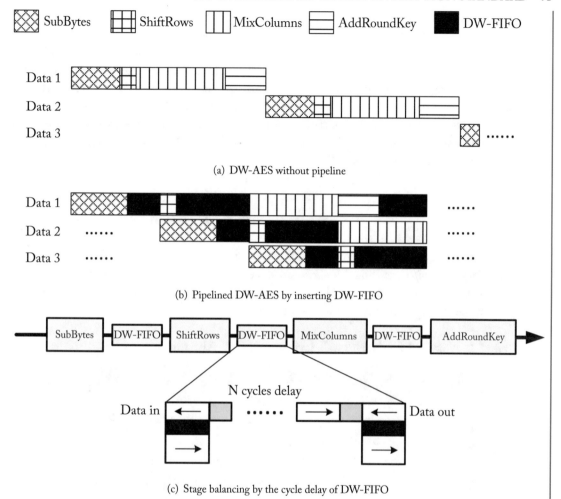

(a) DW-AES without pipeline

(b) Pipelined DW-AES by inserting DW-FIFO

(c) Stage balancing by the cycle delay of DW-FIFO

Figure 3.8: (a) DW-AES without pipelining; (b) pipelined DW-AES by DW-FIFO insertion; and (c) adjusting the latency of DW-FIFOs to balance the pipelined stages.

in Figure 3.8b, the latency of every stage in a pipelined DW-AES can be equalized with the insertion of an appropriate length of DW-FIFO between stages. Thus, the data can be fed in regular interval without requiring sophisticated multirate signal processing technique or multi-clock domain synchronization.

For 128-bit AES, each input data needs to undergo the four stages of operations for ten times. For a four-stage pipeline, the fifth data chunk can only begin after all ten iterations have been completed. This problem can be solved by generating multiple ciphers in series to maintain the same input data rate without the need for extra buffering of input data.

The algorithm for the mapping of DW-AES to a high-speed energy-efficient non-pipelined architecture is described in Algorithm 3.2. The symbols used are listed in Table 3.1. Specifically, it enumerates different amount of resources (e.g., N_{xor} and N_{LUT}) and allocates the resources to each stage (e.g., $\mathcal{R}_{ARK}$ is a set of resources for AddRoundKey stage). Then the delay, power, and area of each stage are evaluated according to each operation detailed in Section 3.1.2. Lastly, the resource configuration with the best target performance (power, area or throughput) will be selected as the final solution $\mathcal{S}_R$ for the hardware performance specified in the design stage. In general, the maximum throughput of DW-AES design without pipelining can be obtained by minimizing the total latency as follows:

$$t_{MinSum} = \underset{\mathcal{S}_R}{\text{minimize}} \quad Sum(t_{SB}, t_{ARK}, t_{SR}, t_{MC}) \tag{3.5}$$
$$\text{subject to} \quad P(\mathcal{S}_R) \le P_m, A(\mathcal{S}_R) \le A_m,$$

where t_{SB}, t_{ARK}, t_{SR}, and t_{MC} are the numbers of cycles required for the execution of *SubBytes*, *AddRoundKey*, *ShiftRows*, and *MixColumns* stages determined by the allocated resources $\mathcal{S}_R$. P_m and A_m are the maximum allowable power and area specified by the design constraints. Consequently, the throughput of non-pipelined DW-AES, $T \propto \frac{1}{t_{MinSum}}$.

Table 3.1: Notations for domain wall device-based DW-AES implementation

Symbols	Descriptions
M_s	4×4 state matrix for the input of 128-bit AES
$S_{i,j}$	state byte, entry in the i-th row and j-th column of M_s
$K_{i,j}$	key bytes correspond to $S_{i,j}$
A_i	the i_{th} DW array for physical implementation of M_s
N_r	number of iterative rounds in AES algorithm
M_{mc}	the MixColumns transformation matrix
T	throughput of AES cipher
N_{LUT}	number of look-up tables
N_{xor}	number of xor units
t	latency of DW-AES measured in number of cycles
$\mathcal{S}_R$	set of resources utilized by DW AES architecture
$\mathcal{R}$	set of resources for each module
M	memory-based logic units in the resource sets

For the pipelined DW-AES architecture, the throughput is limited by the critical stage that has the longest latency. The maximum throughput of pipelined DW-AES design is obtained by minimizing the latency of the most time-critical stage as follows:

$$t_{input} = \begin{array}{c} \text{minimize} \\ S_R \end{array} \quad Max(t_{SB}, t_{ARK}, t_{SR}, t_{MC})$$
$$\text{subject to} \quad P(S_R) \leq P_m, A(S_R) \leq A_m. \tag{3.6}$$

The corresponding throughput of DW-AES is $T_{pipeline} \propto \frac{1}{t_{input}}$. Ideally, if the four stages have the same latency, the throughput of the pipelined DW-AES will be quadrupled. The gain in throughput by pipelining diminishes with increasing latency skew of the critical stage. Taking into account the resource constraint, and limited power and area budgets, the computational resources will need to be relocated from the non-critical stages to the critical stage to improve the latency of the latter. Design space exploration can be performed to find the best pipelined DW-AES configuration with the highest achievable throughput.

Algorithm 3.3 Pipelined DW nanowire-based AES mapping

Require: AES design area constraint A_m or/and power constraint P_m
1: $S_R \leftarrow \emptyset$
2: $\mathcal{R}_{SR} \leftarrow \{32 \times M_{nw}\}$; $\mathcal{R}_{MC} \leftarrow \{4 \times M_{LUT}, 8 \times M_{4-16dec}\}$
3: **for** (N_{xor}, N_{LUT}) combinations, $N_{xor} \in \{1, 2, 4...32\}$ and $N_{LUT} \in \{1, 2, 4\}$ **do**
4:     $\mathcal{R}_{ARK} \leftarrow \{N_{xor} \times M_{xor}, 32 \times M_{sa}\}$
5:     $\mathcal{R}_{SB} \leftarrow \{N_{LUT} \times M_{LUT}, 32 \times M_{sa}, 8 \times M_{4-16dec}\}$
6:     calculate $\{t_{SB}, t_{ARK}, t_{SR}, t_{MC}\} \leftarrow f(\{\mathcal{R}_{SB}, \mathcal{R}_{ARK}, \mathcal{R}_{MC}\})$
7:     $\begin{array}{c} minimize \\ N \in P \end{array} Max\{\frac{t_{SB}}{N_{SB}}, \frac{t_{ARK}}{N_{ARK}}, \frac{t_{SR}}{N_{SR}}, \frac{t_{MC}}{N_{MC}}\}$
8:     $S'_R \leftarrow \{N_{SB} \times \mathcal{R}_{SB}\} \cup \{N_{ARK} \times \mathcal{R}_{ARK}\} \cup \{N_{SR} \times \mathcal{R}_{SR}\} \cup \{N_{MC} \times \mathcal{R}_{MC}\}$
9:     calculate area $A \leftarrow g(S'_R)$, power $P \leftarrow h(S'_R)$ and throughput $T' \leftarrow j(S'_R)$
10:     **if** $A \leq A_m$ and $P \leq P_m$ and $T' \geq T$ **then**
11:         $S_R \leftarrow S'_R$
12:     **end if**
13: **end for**
Ensure: Pipelined DW-AES architecture resource utilization S_R

Multi-issue Pipelined DW-AES

Due to the large disparity in computational complexity of different pipelined stages, resource re-allocation may not be able to adequately reduce the latency skew among all stages. Hence, we propose adopting the multiple issue technique used in super-scalar processor for the pipelined DW-AES. The DW-AES with pipeline/multi-issue mapping algorithm is described in Algorithm 3.3 using the symbols defined in Table 3.1. To overcome the timing bottleneck, additional processing units are introduced to share the workload of any critical stages after the resource reallocation, as shown in Figure 3.9b. In Figure 3.9b, two additional units are added to the bottleneck MixColumns stage. Its timing diagram is shown in Figure 3.9a, which illustrates the throughput improvement of the overall pipelined architecture. In Figure 3.9a, it is observed that the MixColumns stage is twice slower than other stages. Before the application of multiple issue technique, the latency of MixColumns is three time higher than other stages, and the data can only be fed at a rate that is limited by its latency. By adding two more MixColumns units, the MixColumns stage is able to operate at the same speed as other stages, thus the data rate is tripled

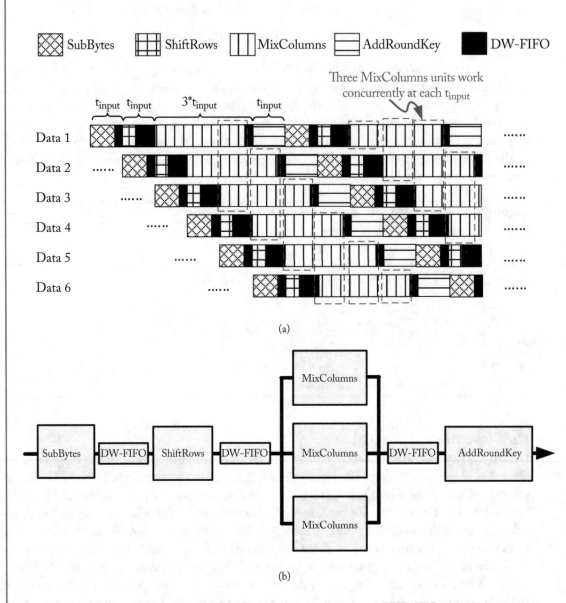

Figure 3.9: (a) Timing diagram and (b) block diagram of pipelined DW-AES with multi-issue.

as the latency of all stages are now equalized. In general, the period of data input clock t_{input} shown in Figure 3.9a can be obtained by

$$t_{input} = t_{MinMax} =$$
$$\underset{S_R}{\text{minimize}} \quad Max(\frac{t_{SB}}{N_{SB}}, \frac{t_{ARK}}{N_{ARK}}, \frac{t_{SR}}{N_{SR}}, \frac{t_{MC}}{N_{MC}}) \qquad (3.7)$$
$$\text{subject to} \quad P(S_R) \leq P_m, A(S_R) \leq A_m,$$

where N_{SB}, N_{ARK}, N_{SR}, and N_{MC} are the number of processing units of *SubBytes*, *AddRoundKey*, *ShiftRows*, and *MixColumns* stages, respectively. The corresponding throughput for the multi-issue pipelined DW-AES is $T_{multiissue} \propto \frac{1}{t_{input}}$.

Compared to the pipelined DW-AES, the multi-issue pipelined DW-AES has a higher throughput. In addition, due to the higher utilization rate, less resources are idle most of the time. Consequently, the ratio of leakage power to total power consumption is lowered, which further improves the energy efficiency of the multi-issue pipelined DW-AES at the expense of area overhead due to the additional resources.

3.1.4 PERFORMANCE EVALUATION

AES Performance Comparison

The proposed DW-AES cipher is compared with both CMOS-based ASIC design [Mathew et al., 2011, 2014] and hybrid CMOS/ReRAM (CMOL) design [Abid et al., 2009]. In ASIC implementation, the performance data is extracted from the reported results in Mathew et al. [2011, 2014] with the following technology scaling: $Area \sim 1/S^2$, $Delay \sim 1/S^2$, $Energy \sim 1/S$, where $S = L/32$ nm and leakage scaling factor from Abbas and Olivieri [2014]. The stacked CMOL AES performance is projected from Abid et al. [2009]. In particular, the lower layer CMOS gates are evaluated with the aforementioned technology scaling together with the interconnection improvement ratio by the upper non-volatile layer. C-code based software implementation that runs on a general purpose processor (GPP) is also compared. The evaluation of the software implementation of AES is done in two steps. First, the AES binary compiled by the C-code obtained from Malbrain [2009] is input to gem5 [Binkert et al., 2011] simulator to generate the run-time utilization rate of core components. Then, the generated statistics are used by McPAT [Li et al., 2009] to produce the core power and area model. Table 3.2 compares the different implementations of AES cipher, and the results are discussed as follows.

Power   As expected, the DW-AES cipher has the smallest leakage power due to the use of non-volatile DW nanowire devices. The small leakage power is mainly introduced by its CMOS peripheral circuits, i.e., decoders, SAs, as well as simple controllers. Specifically, DW-AES cipher reduces the leakage power of CMOS ASIC and memristive CMOL designs by 96% and 98%, respectively. The leakage power can be further reduced if the decoders and SAs of the memory are reused by the DW-AES ciphers. All three types of DW-AES consume comparable energy for the encryption of one 128-bit input block. The slight differences are ascribed to the DW-

Table 3.2: Comparison of different AES implementations on 128-bit encryption performance

Implementation	Leakage (μW)	Energy/128-bit(nJ)	Area (μm^2)	Cycles
C code [Malbrain, 2009] on GPP[†]	1.3e6	460	2.5e+6	2309
CMOS ASIC [Mathew et al., 2014]	400	6.6	4400	336
memristive CMOL [Abid et al., 2009]	624	10.3	320	470
CMOS Pipelined [Mathew et al., 2011]	267-16800[††]	0.9	75000	20
Baseline DW-AES	14.6	2.4	78	1022
Pipelined DW-AES	15.3	2.3	83	2652
Multi-issue DW-AES	21.4	2.7	155	1320

[†] general purpose processor

[††] 267μW@V$_{DD}$=320 mV, and 16.8 mW@V$_{DD}$=1.1V

FIFO and different CMOS-based cycle controllers. Despite consuming slightly higher leakage power than the baseline DW-AES, pipelined DW-AES and multi-issue DW-AES encrypt four input blocks each time, which result in a higher overall energy efficiency. This is because the computational resources of pipelined DW-AES and multi-issue DW-AES are active most of the time. The energy is expended for useful functions than wasted in idleness. The breakdowns of dynamic and leakage power consumptions for each module of all DW based implementations are shown in the bar charts of Figures 3.10c and 3.10d, respectively. The *MixColumns* module that involves more xor and LUT operations consume the majority of the total dynamic power. On the other hand, leakage power is dominated by the *MixColumns* and *SubBytes* modules due to their predominant volatile components such as decoders for DW memory array and SAs.

Area   Benefiting from the high density of DW nanowire devices, DW-AES is highly area-efficient. In particular, substantial areas have been saved by employing very compact DW-LUTs in the two most resource-consuming *SubBytes* and *MixColumns* modules.

As mentioned earlier, the areas reported in Mathew et al. [2011, 2014] and Abid et al. [2009] are technologically scaled to 32 nm technology by a factor of $1/S^2$. The areas of our proposed architectures are calculated based on the number of CMOS transistors, and the sizes of MTJ, DW nanowire and memory array using the DW-CACTI tool developed from CACTI [Wilton and Jouppi, 1996]. As the in-memory architecture is composed of CMOS and DW nanowire circuits, the area consumed in each part is calculated separately. For example, the DW-LUT in Figure 3.3 requires a 256-byte DW memory and two 4-16 decoders. With each do-

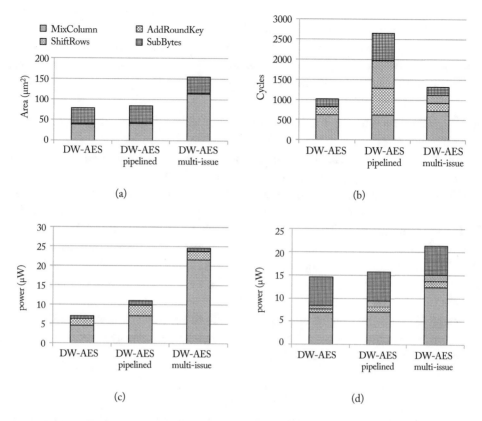

Figure 3.10: The breakdown of (a) area, (b) latency, (c) dynamic power, and (d) leakage power consumed by every module of DW-AES, pipelined DW-AES and multi-issued DW-AES.

main occupying an area of 32 nm × 64 nm, the areas of the DW memory and the two decoders are calculated to be 4.2 μm^2 and 0.95 μm^2, respectively. Because the DW nanowire and CMOS circuits are implemented in different layers, the much smaller CMOS circuit area is fully subsumed under DW nanowire stack. Hence, the SubBytes stage is dominated by the larger nanowire circuit area of 4.2 μm^2. The areas of ShiftRows, AddRoundKey, and MixColumns are calculated in the same way to be 0.04 μm^2, 3.3 μm^2, and 0.3 μm^2, respectively. As the total area of each encryption cipher is only 7.8 μm^2, 10 sets of encryption ciphers are implemented in our design. Pipelined DW-AES and multi-issue DW-AES occupy larger areas than the baseline DW-AES due to the additional register arrays for pipeline and hardware resources like 3 MixColumns for multi-issue.

Overall, the DW-AES cipher saves 98% and 76% of areas over the CMOS ASIC and memristive CMOL designs, respectively. Compared to DW-AES, the pipelined DW-AES incurs a small area overhead of 6.6% due mainly to the DW-FIFO used for stage balancing and additional state matrices. The area of multi-issue DW-AES is almost twice that of DW-AES. This is attributable to the additional processing units. Nevertheless, it still exhibits a noteworthy 52% area reduction compared to CMOL design. The breakdown of areas consumed by different modules of all DW based implementations is shown in Figure 3.10a. Due to the use of DW-LUT, *SubBytes* and *MixColumns* consume almost all of the area.

Latency   The premium for the higher area and energy efficiency of DW-AES is an increase in the number of cycles required for its computation. Multiple-cycle operations of DW-XOR and its DW-LUT are inevitable as these functions need to be preceded by the shift-operation in order to align the target cell with the MTJ. The breakdown of latencies required by different modules for all DW based implementations is shown in Figure 3.10b. Additional cycles are added by DW-FIFO to balance the number of cycles of all pipelined stages for the pipelined DW-AES. The latency of multi-issue DW-AES is only slightly larger than that of DW-AES because stage balancing is augmented by adding processing units, with only a small number of cycles added by DW-FIFO. Although the latency from the input of plaintext to the output of cipher text is longer in pipelined DW-AES and multi-issue DW-AES, they are able to process four times more data than the baseline DW-AES. In other words, the extended latency of these pipelined designs is compensated by a higher processing rate. After all, the overall speed of the cipher is determined by how fast a stream of plaintext data can be encrypted continuously instead of how long it takes to encrypt only one chunk of 128-bit data.

Data Processing Rate and Energy Efficiency Comparison

In what follows, the proposed in-memory DW-AES is compared with other implementations at the system level. For each AES computing platform, the number of AES units is maximized to encrypt the input data stream concurrently subject to a fixed area constraint. Except the proposed in-memory DW-AES, all platforms will incur some I/O energy overhead for data access. Given a 2 mm^2 area design budget, the system configurations for different platforms are summarized in Table 3.3. For more compact implementation such as Mathew et al. [2011], more AES ciphers can be housed within the 2 mm^2 area budget. The memory I/O energy overhead is obtained from CACTI. The processor operates at 2 GHz for software implementation, while all other AES accelerators run at a typical clock-rate of 30 MHz, which achieves the optimal energy efficiency according to Mathew et al. [2011, 2014]. Operating the accelerators at faster clock-rate will adversely affect the system's reliability due to high peak power for densely distributed and concurrent execution. Due to the number of concurrent encryptions within a small footprint, 30 MHz is a reasonable clock rate according to Figure 3.11b for all custom circuit implementations to avoid overheating. Given a uniform operating frequency, the number of cycles for the most time-critical stage is different among all implementations, which is indicated in Table 3.3 for a 128-bit cipher. A figure of merit, called the data processing rate (DPR), is defined to measure the rate of encrypted data within a given area budget for different custom implementations.

Table 3.3: System configurations

AES computing platforms configurations and Date Processing Rate (DPR) under 2 mm² area design budget				
Platforms	**# of AES ciphers**	**Clock-rate**	**Critical stage latency (for 128 bits)**	**DPR (GB/s)**
C code [Malbrain, 2009] on GPP	1 core	2 GHz	2309	0.016
CMOS ASIC [Mathew et al., 2014]	454	30 MHz	84	2.59
Pipelined ASIC [Mathew et al., 2011]	26		5	2.5
memristive CMOL [Abid et al., 2009]	6250		470	6.38
DW-AES	25640		1022	12
DW-AES pipelined	24096		663	17.4
DW-AES multi-issue	12902		220	28
Memory Configurations				
Memory capacity	1GB			
Bus Width	128 bits			
Bandwidth	6.4 GB/s			
I/O energy overhead	3.7nJ per access			

DPR, expressed in bytes per second, is defined as

$$DPR = \frac{\#ciphers \times f \times \#bytes\ per\ cipher}{latency\ of\ critical\ stage}, \tag{3.8}$$

where #ciphers is the number of ciphers that can be implemented in a given area, f is the clock frequency in Hz, #bytes per cipher is the number of bytes in each cipher and the latency of critical stage is the number of clock cycles consumed in the longest stage of AES implementation.

Figure 3.11 compares the DPR, power and energy efficiency of different AES computing platforms. As expected, all AES hardware implementations have several orders of magnitude improvement in DPR and energy efficiency compared to the software implementation on GPP. Among all the hardware implementations, the proposed DW-AES has a DPR of 12 GB/s, which is 4.6× higher than that of the CMOS ASIC-based platform with a similar power consumption, and 1.88× higher than that of the memristive CMOL-based platform with 38% power reduction.

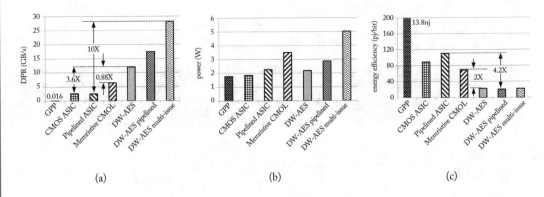

Figure 3.11: Comparison of data processing rate (DPR), power, and energy efficiency of encryption among different in-memory AES computing platforms.

Due to the smaller area per cipher of DW-AES, more ciphers can operate in parallel under the same area budget, leading to its higher DPR. The multi-issue pipelined DW-AES has the highest DPR of 28 GB/s, which is 11× higher than its CMOS ASIC counterpart with 2.2× power penalty. This further improvement in DPR is achieved by reducing the latency of the critical stage. Due to the in-memory encryption and non-volatility, the proposed DW-AES computing platforms have the best energy efficiency of 22pJ/bit, which is 5×, 5.2×, 3× times higher than that of CMOS ASIC, pipelined CMOS ASIC, and memristive CMOL-based platforms, respectively.

 For higher usage of in-memory AES logic, the dynamic power has larger impact on the energy efficiency, while for lower usage of in-memory logic, the leakage power has greater impact. As the DW-AES implementations have comparable energy consumption but much lower leakage power, as shown in Table 3.2, the energy efficiency will increase when the usage rate decreases. For high usage, the advantage of energy efficiency of DW-AES ciphers over ASIC implementations will be somewhat reduced but it remains competitively better owing to its efficient in-memory communication and lower energy consumption per encryption.

3.2   DOMAIN WALL-BASED SIMON BLOCK CIPHER

SIMON is a lightweight cryptographic block cipher introduced by the National Security Agency (NSA) in Ray et al. [2013]. Compared to other block ciphers, SIMON consumes extremely small area, while providing similar level of security, thereby, making it suitable for hardware- and power-constrained applications. SIMON has multiple versions1 based on the size of plaintext word and the key(s). A specific version of SIMON can have plaintext block-sizes of 32, 48, 64, 96, and 128 bits, with up to three key sizes for each block-size.

3.2.1   FUNDAMENTAL OF SIMON BLOCK CIPHER

SIMON 32/64 has a plaintext size of 32-bit. The Boolean expression for SIMON encryption is given as follows:

$$x_{i+2} = S^1 x_{i+1} \cdot S^1 x_{i+1} \oplus x_i \oplus S^2 x_{i+1} \oplus k_i, \qquad (3.9)$$

where x_{i+1} and x_i represent the upper and the lower 16 bits of input plaintext, respectively. Here, S^1, S^2, and S^8 represent left circular shift by 1, 2, and 8 bits, respectively; k_i is the 16-bit key for i-th encryption round; O_1, O_2, O_3, and O_4 are the results of the intermediate logic operations that constitute SIMON encryption-logic. The encryption logic iterates for 32 rounds. The key expansion logic [Ray et al., 2013] for SIMON is given in the following equation:

$$k_{i+m} = c \oplus (z_j)_i \oplus k_i \oplus \left(I \oplus S^{-1} \right) \left(S^{-3} k_{i+3} \oplus k_{i+1} \right). \qquad (3.10)$$

S^{-1} and S^{-3} represent right circular shift by 1 and 3 bits, respectively. Of the 32 keys, the first $4 (= m)$ keys are provided initially. As a result, the key expansion routine needs to be executed for 28 rounds.

3.2.2   HARDWARE STAGES

To reduce power consumption, especially the high energy consumed by the write circuits if they are always ON, the entire circuit-implementation has been divided into eight successive stages. When one stage is ON, the other stages are OFF. A RM-based right circular shifter is used to propagate the ON signal from one stage to another. The racetrack stores 8 bits, and each bit-position in the racetrack corresponds to a stage. Of these 8 bits, only one bit is 0 while the rest are 1s. A stage is set ON when the bit in the racetrack corresponding to that stage is 0. The stage is OFF if the bit is a 1. As a result, at any instant, only one stage is ON. One round of SIMON encryption completes upon the traversal of the "0" from stage 1 to stage 8 in the direction shown in Figure 3.12. Figure 3.13 shows the RM-based implementation of the circular shifter. Each bit-position in the RM has an MTJ sensor which is read by a sensing amplifier (SA). The output of the SA is used in clock gating the circuits in a stage. If the output of an SA is LOW, the corresponding stage is ON, else it is OFF.

3.2.3   ROUND COUNTER

As the key expansion logic and the encryption logic iterate for 28 and 32 rounds respectively, an RM-based counter module has been implemented to count the number of rounds. The round counter in Figure 3.12b comprises of two circular shifters. One circular shifter uses racetrack A to store 32 bits four 1s and twenty-eight 0s. The circular shifting of bits in racetrack A is shown in Figure 3.12b. As a result, the Count 28 signal output of sense amplifier (SA) reading the MTJ on this bit—remains HIGH during rounds 29–32. This marks the completion of 28 rounds. The other circular shifter uses racetrack B to store 4 bits three 1s and a 0. When Count 28 is HIGH, the circular shifting in racetrack B occurs. The sequence of circular shifting of bits in racetrack B

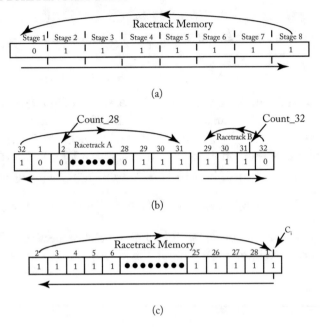

(a)

(b)

(c)

Figure 3.12: Circular shifting of bits through stages.

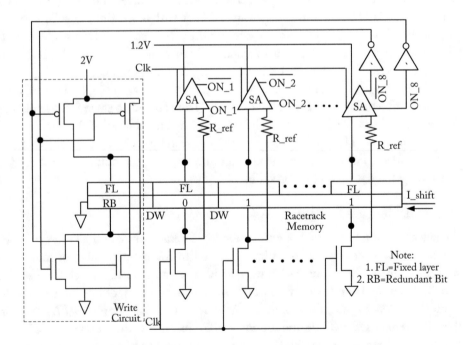

Figure 3.13: Domain wall-based circular shifter.

is shown in Figure 3.12b. In the 8th stage of round 32, the Count 32 signal output of SA reading the MTJ on this bit evaluates to 1 to indicate the completion of 32 rounds.

3.2.4 CONTROL SIGNALS

Figure 3.14 shows the generation of control signals for various stages of the key expansion unit and the encryption unit. Signals $\overline{ON\_i}$ and $\overline{Key\_ON\_i}$ in Figure 3.14 are used for clock gating i-th stage of the encryption unit and the key expansion unit, respectively. When Count_32 is HIGH, output of the encryption unit represents the valid cipher and a new plaintext is input to the circuit for encryption. Circuits in the key expansion logic unit remain switched OFF as long as Count_28 remains HIGH. New 16-bit key is input when the output New_Key_In goes HIGH. New_Key_In goes HIGH when stage 7 is ON during rounds 29–32, so that four new 16-bit keys are input for encryption of the next new plaintext.

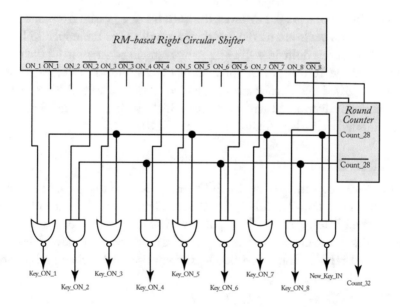

Figure 3.14: Domain wall-based circular shifter.

3.2.5 KEY EXPANSION

For SIMON 32/64, $c = 0xfffc$ and $z_j = z_0$ is a sequence of bits. The first 28 bits (each bit for a round of key-expansion) of this sequence are 1111101000100101011000011100. Since c and z_0 are both constants, $c \oplus (z_0)_i$ is denoted by a constant C_i. Equation 3.10 is re-written as follows:

$$k_{i+m} = C_i \oplus k_i \oplus \left(I \oplus S^{-1}\right)\left(S^{-3}k_{i+3} \oplus k_{i+1}\right). \tag{3.11}$$

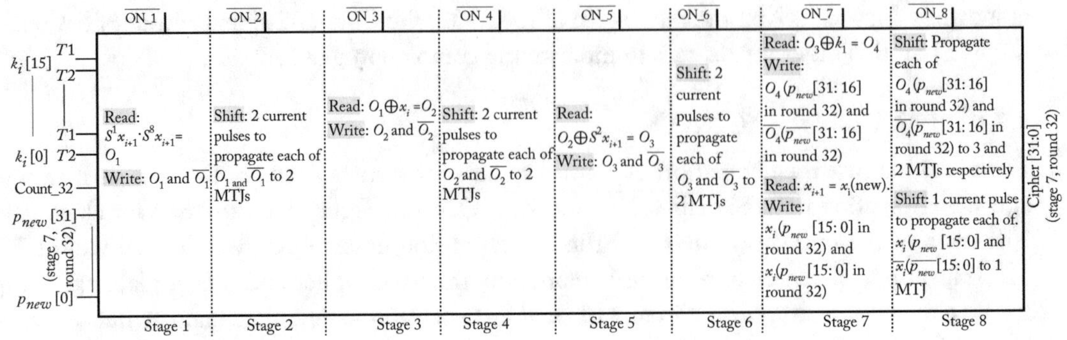

	ON_1	ON_2	ON_3	ON_4	ON_5	ON_6	ON_7	ON_8
$k_i[15]$ T1 T2 $k_i[0]$ T1 T2 Count_32 $p_{new}[31]$ (stage 7, round 32) $p_{new}[0]$	Read: $S^i x_{i+1} \cdot S^8 x_{i+1} = O_1$ Write: O_1 and $\overline{O_1}$	Shift: 2 current pulses to propagate each of O_1 and $\overline{O_1}$ to 2 MTJs	Read: $O_1 \oplus x_i = O_2$ Write: O_2 and $\overline{O_2}$	Shift: 2 current pulses to propagate each of O_2 and $\overline{O_2}$ to 2 MTJs	Read: $O_2 \oplus S^2 x_{i+1} = O_3$ Write: O_3 and $\overline{O_3}$	Shift: 2 current pulses to propagate each of O_3 and $\overline{O_3}$ to 2 MTJs	Read: $O_3 \oplus k_1 = O_4$ Write: $O_4(p_{new}[31:16]$ in round 32) and $\overline{O_4}(\overline{p_{new}}[31:16]$ in round 32) Read: $x_{i+1} = x_i$(new). Write: $x_i(p_{new}[15:0]$ in round 32) and $x_i(\overline{p_{new}}[15:0]$ in round 32)	Shift: Propagate each of $O_4(p_{new}[31:16]$ in round 32) and $\overline{O_4}(\overline{p_{new}}[31:16]$ in round 32) to 3 and 2 MTJs respectively Shift: 1 current pulse to propagate each of. $x_i(p_{new}[15:0]$ and $\overline{x_i}(\overline{p_{new}}[15:0]$ to 1 MTJ	Cipher[31:0] (stage 7, round 32)
	Stage 1	Stage 2	Stage 3	Stage 4	Stage 5	Stage 6	Stage 7	Stage 8	

Figure 3.15: SIMON Encryption unit.

In Equation 3.11, C is 16-bit long, while z_0 is a sequence of bits. The first 14 bits of C_i are 1s, and the 15th bit of C_i is 0. However, the 16th bit of C_i varies from one round to another. Therefore, two domain walls are used. One domain wall stores the upper 15 bits of C_i and $\overline{C_i}$. No shifting is done in this domain wall as these bits remain constant in all the rounds. The other domain wall stores the 16th bit of C_i for each of the 28 rounds of key-expansion. The bits in this racetrack are shifted in a circular manner, which continues up to 28th round. The entire key expansion execution is distributed across 8 stages, which can iterate up to 28 rounds. All the logic operations performed in the key expansion unit are XORs.

3.2.6   ENCRYPTION

A detailed stage-wise depiction of the SIMON encryption unit is given in Figure 3.15. The internal logic outputs O_1, O_2, O_3, and O_4 depicted in Figure 3.16 are read and written to an MTJ in stages 1, 3, 5, and 7, respectively. The input key-bits $k_i[15]...k_i[0]$ for i-th round of encryption are outputs of the key-expansion module. Cipher bits are output and new plaintext bits are input when $Count_{32}$ goes HIGH. The encryption logic comprises of XOR and AND operations.

3.2.7   PERFORMANCE EVALUATION

ST Microelectronics 65 nm PDK and the SPINLIB Verilog-A compact models for Perpendicular Magnetic Anisotropy RM [Zhang et al., 2012b] and MTJ [Zhang et al., 2012a] have been used to simulate the CMOS-RM hybrid circuit in Cadence Virtuoso Analog Design Environment simulator. The models are based on current induced Spin Torque Transfer (STT) phenomenon for switching of CoFeB/MgO/CoFeB MTJ and shifting of DWs in the CoFeB racetrack. Values of some of the important parameters of the circuit are listed in Table 3.4. We have chosen the amplitude of all shift pulses in the implementation to be 176 μA [Zhang et al., 2012b]. The pulse width corresponding to this amplitude is 0.8 ns. In stage 2, 3 such pulses are applied to the racetrack with a gap of 0.1 ns between each pulse. As a result, the time period of clock cycle is

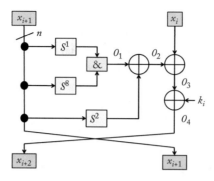

Figure 3.16: SIMON encryption scheme.

Table 3.4: Simulation parameters

Parameters	Descriptions	Values
V_{write}	Writing voltage	2 V
W_{write}	Width of transistors in writing circuits	0.6 μm
V_{read}	Reading voltage	1.2 V
W_{read}	Width of transistors in read circuits	0.135 μm
I_{shift}	Amplitude of shifting pulses	176 μA

Table 3.5: Encryption energy of 2-input logic gate based implementation

Component	Energy (nJ)
Writing	25.62
Domain-wall shifting	2.82
Read + CMOS gates	0.08
Total Energy	28.52

2.7 ns. As the width of the CMOS transistors is reduced, writing delay increases. Using a large width of transistor may result in a writing delay less than 2.7 ns. This will result in the energy-hungry write circuits being ON for surplus amount of time leading to a wastage of energy. The minimum width that can successfully write into MTJ within 2.7 ns is 0.6 μm. Therefore, we have set the transistor-widths at this value.

The 2-input logic gate based implementation of SIMON operates at a frequency of 370.37 MHz. The energy consumed in encrypting one plaintext and the constituents of this computation energy are shown in the following Table 3.5. Table 3.5 shows a comparison of the energies consumed by read, write, and shift operations. Almost 90% of the computation energy is consumed by write operations. Domain wall nanowire shifting consumes about 9.8% of the total computation energy. Energy consumed by the CMOS read circuits and logic gates is negligible in comparison to the write and shift energies. While benchmarking this result with comparable implementation of SIMON32/64 in 65 nm CMOS, we observe that, our SIMON implementation is two orders of magnitude worse in terms of energy consumption. This forms the motivation to study potential design optimizations for us.

3.3 REFERENCES

Abbas, Z. and Olivieri, M. (2014) Impact of technology scaling on leakage power in nanoscale bulk CMOS digital standard cells. *Microelectronics Journal*, 45(2), pp. 179–195. DOI: 10.1016/j.mejo.2013.10.013. 97

Abid, Z. et al. (2009) Efficient CMOL gate designs for cryptography applications. *IEEE Transactions in Nanotechnology*, 8(3), pp. 315–321. DOI: 10.1109/tnano.2008.2011812. 82, 97, 98

Augustine, C. et al. (2011) Numerical analysis of domain wall propagation for dense memory arrays. In *Proc. 2011 IEEE Int. Electron Devices Meeting*, Washington, DC, pp. 17.6.1–17.6.4. DOI: 10.1109/iedm.2011.6131575. 87

Bertoni, G., Breveglieri, L., Koren, I., Maistri, P., and Piuri, V. (2003) Error analysis and detection procedures for a hardware implementation of the advanced encryption standard. *Computers, IEEE Transactions on*, 52(4), pp. 492–505. DOI: 10.1109/tc.2003.1190590. 92

Binkert, N. et al. (2011) The gem5 simulator. *ACM SIGARCH Computer Architecture News*, 39(2), pp. 1–7. DOI: 10.1145/2024716.2024718. 97

Daemen, J. and Rijmen, V. (2002) *The Design of Rijndael: AES—The Advanced Encryption Standard*, Springer, Berlin, Germany. DOI: 10.1007/978-3-662-04722-4. 90

Karri, R., Wu, K., Mishra, P., and Kim, Y. (2001) Fault-based side-channel cryptanalysis tolerant rijndael symmetric block cipher architecture. In Defect and Fault Tolerance in VLSI Systems, *Proceedings 2001 IEEE International Symposium*, pp. 427–435. DOI: 10.1109/dftvs.2001.966796. 92

Li, S. et al. (2009) McPAT: an integrated power, area, and timing modeling framework for multicore and manycore architectures. In *Proc. 2009 IEEE/ACM Int. Symp. Microarchitecture*, New York, pp. 469–480. DOI: 10.1145/1669112.1669172. 97

Malbrain, K. (2009) Byte-oriented-aes: A public domain byte-oriented implementation of AES in C. https://code.google.com/p/byte-oriented-aes/ 97

Mathew, S. et al. (2011) 53 Gbps native GF$(2^4)^2$ composite-field AES-encrypt/decrypt accelerator for content-protection in 45 nm high-performance microprocessors. *IEEE Journal of Solid-State Circuits*, 46(4), pp. 767–776. DOI: 10.1109/vlsic.2010.5560310. 97, 98, 100

Mathew, S. et al. (2014) 340mV-1.1V, 289Gbps/W, 2090-gate NanoAES hardware accelerator with area-optimized encrypt/decrypt $GF(2^4)^2$ polynomials in 22 nm tri-gate CMOS. In *Proc. 2014 Symp. VLSI Circuits*, Honolulu, HI, pp. 1–2. DOI: 10.1109/vlsic.2014.6858420. 97, 98, 100

Nogami, Y., Nekado, K., Toyota, T., Hongo, N., and Morikawa, Y. (2010) Mixed bases for efficient inversion in $\mathbf{F}((2^2)^2)2$ and conversion matrices of subbytes of AES. In *Cryptographic Hardware and Embedded Systems, CHES 2010*, Springer, pp. 234–247. DOI: 10.1007/978-3-642-15031-9_16. 85

Ray, B., Douglas, S., Jason, S., Stefan, T., Bryan, W., and Louis, W. (2013) The simon and speck families of lightweight block ciphers, Technical Report, Cryptology ePrint Archive, Report./404 102, 103

Satoh, A., Sugawara, T., and Aoki, T. (2009) High-performance hardware architectures for galois counter mode. *Computers, IEEE Transactions on*, 58(7), pp. 917–930. DOI: 10.1109/tc.2008.217. 91

Usselmann, R. (2002) Advanced Encryption Standard/Rijndael IP Core. `http://opencores.org/project,aes_core` 81

Wang, Y., Ni, L., Chang, C. H., and Yu, H. (2016) Dw-aes: A domain wall nanowire based AES for high throughput and energy-efficient data encryption in non-volatile memory. *IEEE Transactions on Information Forensics and Security*, pp. 1–1. DOI: 10.1109/tifs.2016.2576903. 83

Wilton, S. and Jouppi, N. (1996) CACTI: an enhanced cache access and cycle time model. *IEEE Journal of Solid-State Circuits*, 31(5), pp. 677–688. DOI: 10.1109/4.509850. 98

Zhang, Y., Zhao, W., Lakys, Y., Klein, J. O., Kim, J. V., Ravelosona, D., and Chappert, C. (2012a) Compact modeling of perpendicular-anisotropy cofeb/mgo magnetic tunnel junctions. *IEEE Transactions on Electron Devices*, 59(3), pp. 819–826. DOI: 10.1109/ted.2011.2178416. 106

Zhang, Y., Zhao, W., Ravelosona, D., Klein, J. O., Kim, J., and Chappert, C. (2012b) Perpendicular-magnetic-anisotropy cofeb racetrack memory. *Journal of Applied Physics*, 111(9), pp. 093,925. DOI: 10.1063/1.4716460. 106

CHAPTER 4

In-memory Data Analytics

Abstract   The analysis of big-data at exascale (10^{18} bytes/s or flops) has introduced the emerging need to reexamine the existing hardware platform that can support memory-oriented computing. A big-data-driven application requires huge bandwidth with maintained low-power density. The most widely existed data-driven application is machine learning in big data storage system, as the most exciting feature of future big-data storage system is to find implicit pattern of data and excavate valued behavior behind. However, to handle big image data at exascale, there is a memory wall that has long memory access latency as well as limited memory bandwidth. As the third-generation of spin-based memory, domain wall nanowire, also known as racetrack memory [Parkin and et al., 2008, Thomas et al., 2011], is a newly introduced NVM device that can have multiple bits densely packed in one single nanowire, where each bit can be accessed by the manipulation of the domain wall. Compared with STT-RAM, the domain wall nanowire is able to provide the similar speed and power but with much higher density or throughput [Venkatesan et al., 2012]. Since domain wall nanowire has close-to-DRAM density but with close-to-zero standby power, it becomes an ideal candidate for future main memory that can be utilized for big-data processing. In this chapter, we will introduce how to utilize domain wall nanowire on machine learning for image processing. Sparse-representation-based face recognition is also accelerated on this platform.

4.1   IN-MEMORY MACHINE LEARNING

4.1.1   EXTREME LEARNING MACHINE

Among numerous machine learning algorithms [Furey et al., 2000, Hagan et al., 1996, Huang et al., 2006, Suykens and Vandewalle, 1999, Yegnanarayana, 2004], support vector machine (SVM) [Furey et al., 2000, Suykens and Vandewalle, 1999] and neural network (NN) [Hagan et al., 1996, Yegnanarayana, 2004] are widely discussed. However, both two algorithms have major challenging issues in terms of slow learning speed, trivial human intervene (parameter tuning) and poor computational scalability [Huang et al., 2006]. Extreme Learning Machine (ELM) was initially proposed [Huang et al., 2004, 2006] for the single-hidden-layer feed-forward neural networks (SLFNs). Compared with traditional neural networks, ELM eliminates the need of parameter tuning in the training stage and hence reduces the training time significantly. The output function of ELM is formulated as (only one output node is considered)

$$f_L = \sum_{i=1}^{L} \beta_i h_i(X) = h(X)\beta, \qquad (4.1)$$

where $\beta = [\beta_1, \beta_2, \cdots, \beta_L]^T$ is the output weight vector storing the output weights between the hidden layer and output node. $h(X) = [h_1(X), h_2(X), \cdots, h_L(X)]^T$ is the hidden layer output matrix given input vector X and performs the transformation of input vector into L-dimensional feature space. The training process of ELM aims to obtain output weight vector β and minimize the training error as well as the norm of output weight

$$Minimize : \|H\beta - T\| \text{ and } \|\beta\| \qquad (4.2)$$

$$\beta = H^\dagger T, \qquad (4.3)$$

where $H^\dagger$ is the Moore-Penrose generalized inverse of matrix H.

Note that ELM-SR is commonly used as pre-processing stage to improve image quality before applying other image algorithms. It involves intensive matrix operation, such as matrix addition, matrix multiplication, as well as exponentiation on each element of a matrix. Figure 4.1 illustrates the computation flow of ELM-SR, where input vector obtained from input image is multiplied by input weight matrix. The result is then added with bias vector b to generate the input of sigmoid function. Lastly sigmoid function outputs are multiplied with output weight matrix to produce final results.

4.1.2   MAPREDUCE-BASED MATRIX MULTIPLICATION

MapReduce Dean and Ghemawat [2008] is a parallel programming model to efficiently handle large volume of data. The idea behind MapReduce is to break down large task into multiple sub-tasks, and each sub-task can be independently processed by different *Mapper* computing units, where intermediate results are emitted. The intermediate results are then merged together to form the global results of the original task by the *Reducer* computing units.

The problem to solve is $x = M \times v$. Suppose M is an $n \times n$ matrix, with element in row i and column j denoted by m_{ij}, and v is a vector with length of n. Hence, the product vector x also has the length of n, and can be calculated by

$$x_i = \sum_{j=1}^{n} m_{ij} v_j = \sum_{j=1}^{n} \sum_{k=1}^{l} b_{ijk}, $$

where the multiplication of $m_{ij} v_j$ is decomposed into the sum of b_{ijk}. As such, the matrix multiplication can be purely calculated by addition operations, and thus the domain wall adder logic can be exploited.

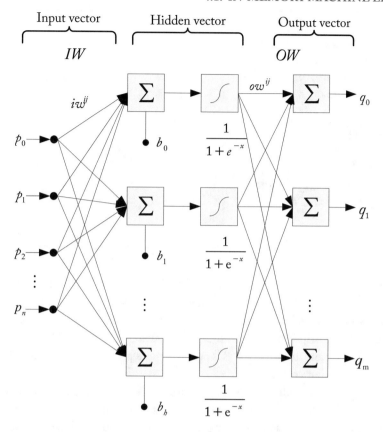

Figure 4.1: The working flow of extreme learning machine.

The pseudo-code of matrix multiplication in MapReduce form is demonstrated in Algorithm 4.1. Matrix M is partitioned into many blocks, and each Mapper function will take the entire vector v and one block of matrix M. For each matrix element m_{ij}, it decomposes multiplication of $m_{ij}v_j$ into additions of multiple b_{ijk} and emits the key-value pair (i, b_{ijk}). The sum of all the values with same key i will make up the matrix-vector product element x_i. A reducer function simply has to sum all the values associated with a given key i. The summation process can be executed concurrently by iteratively summing two values and emitting one result until only one key-value pair is left for each key, namely the (i, x_i).

4.1.3   DOMAIN WALL-BASED HARDWARE MAPPING

Figure 4.2 shows how the MapReduce-based ELM-SR is mapped into the proposed non-volatile memory-based computing platform [Wang et al., 2015]. It is one specific implementation of the local in-memory logic and data pair, as shown in Figure 1.20. The execution starts with a command

---

Algorithm 4.1 Matrix multiplication in MapReduce form

Function: Mapper *partitioned matrix* $p \in M, v$
For *elements* $m_{ij} \in p$
$b_{ijk} \leftarrow decompose(m_{ij}v_j)$
emit(i, b_{ijk}) *to list* l_i
End For
End Function
Function: Reducer l_q
If *length of* $l_q > 1$
remove $(q, v_1), (q, v_2)$ *from list* l_i
emit$(q, v_1 + v_2)$ *to list* l_i
End If
End Function

---

issued by external processor to the memory. The local controller in the in-memory logic part, a simple state machine for example, then loads data from the data array: the off-line trained input weight matrix M and vector v that represents the low-resolution image. A *map* process follows to decompose the multiplications into multiple values by domain wall shift operations, and then emit <key, value> pairs accordingly. All emitted pairs are stored in a separated segment of data array called intermediate results pool.

The <key, value> pairs are further combined in the *reduce* process. Specifically, the controller will fetch elements in the intermediate results pool and dispatch them to available reducers, namely domain wall adders as introduced in Chapter 2. Each reducer will take two values with same key, combine the values by addition, and emit a new pair to the intermediate results pool. The reduce process works in an iterative manner, combining two pairs to one pair until the intermediate results cannot be further combined. Instead of the single addition latency of domain wall adder, the parallelism of additions in MapReduce fashion has direct impact on the latency of obtaining weighted sum results, especially in the data-intensive machine learning application. Therefore, although the domain wall based adder needs multiple cycles to execute, this disadvantage can be suppressed in the MapReduce matrix multiplication.

The domain wall LUTs, configured to execute sigmoid function, are used to form the hidden layer vector. The hidden layer vector then multiplied by the output weight matrix, which is omitted in Figure 4.2 for simplicity of illustration, as it is a repetition of steps 2–4. Sigmoid function includes exponentiation, division, and addition, which is a computing-intensive operation in ELM application. In particular, the exponentiation will take many cycles to execute in the conventional processor due to the lack of corresponding accelerator. Therefore, it is extremely economical to perform exponentiation by look-up table. Note that the LUT size is determined by the input domain, the output range, and the required precision for the floating point numbers. Figure 4.3 shows the ideal logistic curve and approximated curves by LUTs. It can be observed

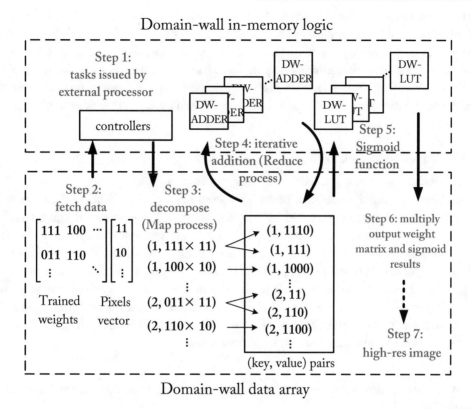

Figure 4.2: ELM-SR algorithm mapping to proposed domain wall nanowire-based computing platform.

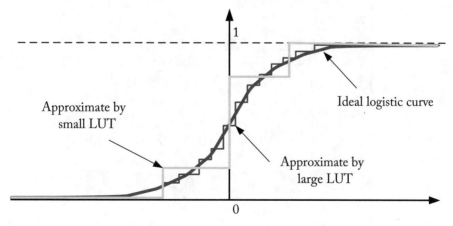

Figure 4.3: Domain wall LUT size effect on the precision of the sigmoid function. The larger the LUT, the smoother and more precise the curve is.

that the output range is bounded between 0 and 1, and although the input domain is infinite, it is only informative in the center around 0. The LUT visually is the digitalized logistic curve and the granularity, i.e., precision, depends on the LUT size. For machine learning application, the precision is not as sensitive as scientific computations. As a result, the LUT size for sigmoid function can be greatly optimized and leads to high energy efficiency for sigmoid function execution. As such, the final results are obtained, and the results written back signifies the end of whole process.

4.1.4   PERFORMANCE EVALUATION

System Level Evaluation of Domain wall Logic and Memory

For machine learning applications, besides the core matrix multiplication there are still many instructions that cannot be executed by MapReduce-based in-memory domain wall logic. In this part, both the domain wall nanowire-based external processor and memory (as shown in Figure 1.20) are evaluated. The core and memory configurations are shown in Table 4.1. The conventional logic in external processor is replaced by their domain wall logic and memory counterparts. The 32-bit 65 nm processor is assumed with four cores integrated. In each core, there are six integer ALUs which execute XOR, OR, AND, NOT, ADD, and SUB operations, and complex integer operations like MUL, DIV are executed in integer MUL. The 32 nm technology node and 64-bit I/O bus width are assumed for memory.

Table 4.1: External processor and memory configurations for real case domain wall logic and memory evaluation

Processor	
Number of cores	4
Frequency	1GHz
Architecture	x86, O3, issue width - 4, 32 bits
Functional units	Integer ALU - 6 Complex ALU - 1 Floating point unit - 2
Cache	L1: 32 KB - 8 way/32 KB - 8 way L2: 1 MB - 8 way Line size - 64 bytes
Memory	
Technology node	32 nm
Memory size	2 GB - 128 MB per bank
IO bus width	64 bits

For domain wall logic based ALU design evaluation, firstly the gem5 simulator is employed to take both SPEC2000 and Phoenix benchmarks [Ranger and et al., 2007] and to generate the runtime instruction and memory accessing traces. The trace file is then analyzed with the statistics of instructions that can be executed on the proposed XOR and adder for logic evaluation. The L2-cache-miss rates are also generated, in order to obtain the actual memory access for memory power evaluation. Then, McPAT is modified to evaluate power of the 32-bit ALU that is able to perform XOR, OR, AND, NOT, ADD and SUB operations. The instruction controlling decoder circuit is also considered during the power evaluation. The leakage power of both designs is calculated at gate level by the McPAT power model.

Figure 4.4 presents the per-core ALU power comparison between the conventional CMOS design and domain wall logic based design. Benefited from the use of domain wall logic, both of the dynamic power and leakage power can be greatly reduced. It can be observed that the set of Phoenix benchmarks consume higher dynamic power compared to those of SPEC2006, which is due to the high parallelism of MapReduce framework with high utilization rate of the ALUs. Among each set, the power results exhibit a low sensitivity to the input, which indicates that percentages of instructions executed in XOR and ADDER of ALU are relatively stable even for different benchmarks. The stable improvement ensures the extension of the proposed domain wall logic to other applications. Averagely, a dynamic power reduction of 31% and leakage power reduction of 65% can be achieved for ALU logic based on all the eight benchmarks.

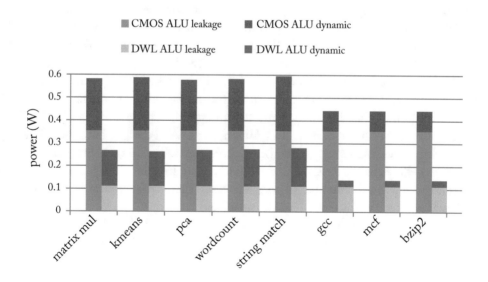

Figure 4.4: The per core ALU power comparison between CMOS design and domain wall logic-based design.

Table 4.2 shows the 128 MB memory-bank comparison between CMOS-based memory (or DRAM) and domain wall nanowire based memory (or DWM). The number of access ports in main memory is varied for design exploration. The results of DRAM are generated by configuring the original CACTI with 32 nm technology node, 64-bit of I/O bus width with leakage optimized. The results of the DWM are obtained by the modified CACTI.

Table 4.2: Performance comparison of 128 MB memory-bank implemented by different structures

Memory Structure	Area (mm^2)	Access Energy (nJ)	Access Time (ns)	Leakage (mW)
DRAM	20.5	0.77	3.46	620.2
DWM/1 port	8.9	0.65	1.90	48.4
DWM/2 ports	6.2	0.72	1.71	30.1
DWM/4 ports	6.2	0.89	1.69	24.3
DWM/8 ports	5.7	1.31	1.88	19.0

It can be observed that the memory area is greatly reduced in the DWM designs. Specifically, the DWMs with 1/2/4/8 access ports can achieve the area saving of 57%, 70%, 70%, and 72%, respectively. The trend also indicates that the increase of number of access-ports will lead to higher area saving. Note that the area saving in turn results in a smaller access latency, and hence the DWM designs on average provide 1.9X improvement on the access latency. However, the DWM needs one more cycle to perform shift operation, which will cancel out the latency advantage. Overall, the DWM and DRAM have similar speed performance. In terms of power, the DWM designs also exhibit benefit with significantly leakage power reduction. The designs with 1/2/4/8 access ports can achieve 92%, 95%, 96%, and 97% leakage power reduction rates, respectively. The advantage mainly comes from the non-volatility of domain wall nanowire-based memory cells. The reduction in area and decoding peripheral circuits can further help leakage power reduction in DWM designs. In addition, the DWM designs have the following trend of access energy when increasing the number of access ports. The designs with 1/2 ports require 16% and 6% less energy, while designs with 4/8 ports incur 15% and 70% higher access energy cost. This is because when the number of ports increases, there are more transistors connected to the bit-line. It leads to the increased bit-line capacitance.

The runtime dynamic power comparison under different benchmark programs are shown in Figure 4.5a. It can be seen that the dynamic power is very sensitive to the input benchmark, and the results of the Phoenix benchmarks show no significant difference from those in SPEC2006. This is because the dynamic power is effected by both intended memory access frequency and the cache miss rate. Although the data-driven Phoenix benchmarks have much higher intended memory reference rate, the miss rate of L1 and L2 cache in Phoenix benchmarks are much lower than SPEC2006, which is due to the very predictable memory access pattern when exhaustively

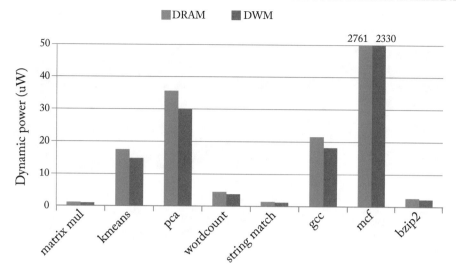

Figure 4.5: (a) The runtime dynamic power of both DRAM and DWM under Phoenix and SPEC2006, and (b) the normalized intended memory accesses.

handling data in Phoenix benchmarks. Overall, the low cache miss rates of Phoenix benchmarks cancel out the higher memory reference demands, which leads to a modest dynamic power. Also, the runtime dynamic power contributes much less to the total power consumption compared to leakage power, thus the leakage reduction should be the main design objective when determining the number of access ports.

Architecture Level Evaluation of Distributed In-memory Computing

In this part, we will show the throughput and energy efficiency improvement brought by proposed purely domain wall nanowire based distributed in-memory computing architecture. As a case study, the data-driven extreme learning machine based super-resolution (ELM-SR) application is executed within the proposed architecture. We will compare the proposed in-memory platform with the conventional general purpose processor- (GPP) based platform. The evaluation of ELM-SR in GPP platform is based on gem5 and McPAT for core power and area model. Proposed in-memory computing architecture is evaluated in our developed self-consistent simulation platform based on NVMSPICE, DW-CACTI, and ELM-SR behavioral simulator. The processor runs at 3GHz while the accelerators run at 500 MHz. System memory capacity is set as 1 GB, and bus width is set as 128 bits. Based on most recent on-chip interconnect and PCB interconnect studies in Park et al. [2013] and Kumar et al. [2014], 40 fJ/bit/mm for on-chip interconnect and 30 pJ/bit/cm for PCB interconnect are used as I/O overhead. For core-memory distance, 10 mm

Table 4.3: Area, power, throughput, and energy efficiency comparison between in-memory architecture and conventional architecture for ELM-SR

Platform	Proposed	GPP (on-chip memory)	GPP (off-chip memory)
# of logic units	1 × processor ·7714 × DW-ADDER 551 × DW-LUT 1 × controller	1 × processor	1 × processor
Logic Area (mm^2)	18 (processor) + 0.5 (accelerators)	18	18
Logic Power (Watt)	10.1	12.5	12.5
Throughput (MBytes/s)	108	9.3	9.3
EPB (nJ)	7	Total: 394 I/O: 364 (92%) Logic: 30 (8%)	Total: 4127 I/O: 4097 (99%) Logic: 30 (1%)

is assumed for on-chip case and 10 cm is assumed for PCB trace length, both according to Park et al. [2013] and Kumar et al. [2014].

Table 4.3 compares ELM-SR in both proposed in-memory computing platform and GPP platform. Due to the deployment of in-memory accelerators and high data parallelism, the throughput of proposed in-memory computing platform is improved by 11.6x compared to GPP platform. In terms of area used by computational resources, proposed in-memory computing platform is 2.7% higher than that of GPP platform. Additional 0.5 mm^2 is used to deploy the domain wall nanowire based accelerators. Thanks to the high integration density of domain wall nanowires, the numerous accelerators are brought with only slight area overhead. In proposed in-memory computing platform, the additional power consumed by accelerators is compensated by the saved dynamic power of processor, since the computation is mostly performed by the in-memory logic. Overall, proposed in-memory computing platform achieves a power reduction of 19%. The most noticeable advantage of proposed in-memory computing platform is its much higher energy efficiency, energy-per-bit (EPB) as metrics, compared to GPP. Specifically, it is 56x and 590x better than that of GPP with on-chip and off-chip memory respectively. The advantage comes from three aspects: (a) in-memory computing architecture that saves I/O overhead; (b) non-volatile domain wall nanowire devices that are leakage free; and (c) application specific ac-

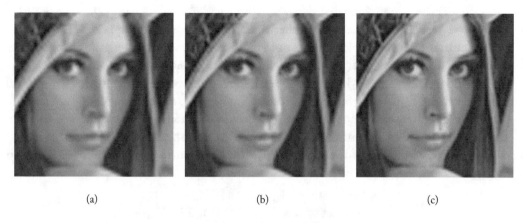

(a)   (b)   (c)

Figure 4.6: (a) Original image before ELM-SR algorithm (SSIM value is 0.91); (b) image quality improved after ELM-SR algorithm by DW-NN hardware implementation (SSIM value is 0.94); and (c) image quality improved by GPP platform (SSIM value is 0.97).

celerators. Specifically, the use of domain wall logic/accelerators contributes to 4x improvement, while the in-memory architecture contributes to the rest (save of I/O overhead).

Figure 4.6 shows the image quality comparison between the proposed in-memory architecture hardware implementation and the conventional GPP software implementation. To measure the performance quantitatively, structural similarity (SSIM) [Wang et al., 2004] is used to measure image quality after ELM-SR algorithm. It can be observed that the images after ELM-SR algorithm in both platforms have higher image quality than the original low-resolution image. However, due to the use of LUT, which trades off precision against the hardware complexity, the image quality in DW-NN is slightly lower than that in GPP. Specifically, the SSIM is 0.94 for DW-NN, 3% lower than 0.97 for GPP.

4.2   IN-MEMORY FACE RECOGNITION

4.2.1   ENERGY-EFFICIENT STT-MRAM WITH SPARE-REPRESENTED DATA

In this section, we will first discuss the sparse-representation based data analytics. Furthermore, we show that the solution of such sparse-represented data maintains the same under random projection and robust to noise, which allows us to partition the data based on required quality of service (QoS).

Sparse-representation based Data Analytics
Sparse-representation is greatly linked to compressed sensing. Under compressed sensing, the signal is directly acquired based on the contents of the signal other than Nyquist-sampling fre-

quency under the assumptions that signal is a sparse under a known basis [Donoho, 2006]. However, sparse-representation data go further than that. Instead of using traditional method such as Fourier transform or wavelet transform to find the basis, we train a matrix as basis, where the interest signal is sparse under such basis, which is also called dictionary learning [Mairal et al., 2009]. Under such framework, we can use sparse-represented data to represent the original data and stores them. Depending on the application required, we can further process for classification or recognition problem or simply recover the data by L_1-norm. By exploring the L_1-norm, we also find the solution of L_1-norm is insensitive to dimension reduction and noise, which greatly relieves the complexity of hardware implement. Therefore, such L_1-norm problem can be formulated as

$$\arg \min \|X\|_0$$
$$\text{subject to } R \times Y = R \times D \times X + Er_{NVM}, \tag{4.4}$$

where $R \in \mathbb{R}^{k \times d}$ is the random projection matrix, $Y \in \mathbb{R}^d$ is the test data and $D \in \mathbb{R}^{d \times t}$ is the training matrix for the specific applications. Er_{NVM} is the error occurred in data storage. Details on each parameter will be further discussed in the following sections. Such L_0-norm optimization problem is equivalent to L_1-norm and can be explained as Figure 4.7a and Figure 4.7b also shows that why L_2-norm provides a different solution from L_0-norm.

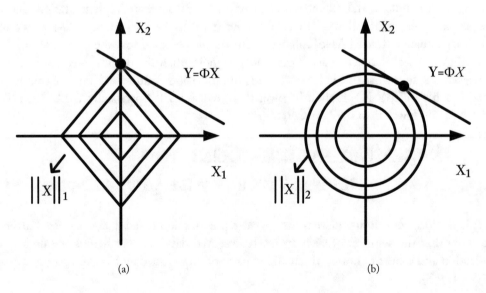

(a)   (b)

Figure 4.7: (a) L_1 norm solution in 2D example, and (b) L_2 norm solution in 2D example.

Random Projection for Spare-representation

In random projection, the original high dimensional data is projected onto a lower-dimensional sub-space using a random matrix with unit column values. Unlike principle component analysis (PCA), random projection is a computationally efficient, yet provides sufficient accuracy for data reduction as,

$$D'_{k \times t} = R_{k \times d} \times D_{d \times t}. \tag{4.5}$$

The dimension reduction from random projection $R \in \mathbb{R}^{k \times d}$ is defined as $\gamma = 1 - \frac{k}{d}$ in sparse representation, where γ is the dimension reduction ratio with $0 < \gamma \leq 1$. The complexity for forming the random matrix R and projecting to $k \times n$ data matrix is of order $O(dkn)$.

The key idea of such random projection is shown in Johnson-Lindenstrauss lemma. This lemma can be applied to show that for an arbitrary matrix $A \in \mathbb{R}^{d \times n}$, given any $\epsilon \geq 0$, there is a mapping $f : \mathbb{R}^d \mapsto \mathbb{R}^k$ for any $k \geq \frac{12 \log(n)}{\epsilon^2}$ such that any two rows $u, v \in A$, we have

$$\begin{aligned}
(1 - \epsilon) \| f(u) - f(v) \|^2 &\leq \| u - v \|^2 \\
&\leq (1 + \epsilon) \| f(u) - f(v) \|^2,
\end{aligned} \tag{4.6}$$

where we can reduce the dimension to $k = O(\frac{\log(n)}{\epsilon^2})$ as long as we can preserve the distance up to a factor of $1 \pm \epsilon$.

Strictly speaking, Equation 4.5 is not a projection since R is not orthogonal. Furthermore, a linear mapping as Equation 4.5 may cause significant distortions. Luckily, by applying the result from Hecht-Nielsen [1994], in a high dimensional space, there exists a much large number of almost orthogonal than orthogonal directions. Thus, random generated matrix might be sufficiently close to orthogonal which reduces the distortion to the original data. Therefore, we will first find the maximum dimension reduction while maintain the property of almost orthogonal and then identify the maximum tolerated error for Er_{NVM}. Data within this limit can be regarded as non-critical data for data-partitions.

Data Partition under Sparse-representation

Data partition is a widely used technique in memory storage due to the fact that typical applications including games, audio/video processing are insensitive to errors but a small portion of the overall of their data [Liu et al., 2012, Venkatesan et al., 2006]. Such data is called critical data. Our proposed sparse-representation data can further utilize this fact to save the power for data storage. Please note that the solution to L_1-norm is robust for bounded noise for classification [Wright et al., 2009]. This is due to that bounded noise does not destroy the largest correlation of input data Y and basis D. However, for data recovery from X to Y, recovery is stable and obtaining a reconstruction with an error term whose size is guaranteed to be proportional to the noise level, which is the best one we can hope [Candes et al., 2006]. The best noise level we can get given that T_0 is the active set of Dictionary D:

$$\hat{X} - X_0 = (D_{T_0}^* D_{T_0})^{-1} D_{T_0}^* Er_{NVM}, \tag{4.7}$$

where $\hat{X}$ is the sparse-representation data from Equation 4.4 and X_0 is the ideal sparse represen-tation data without noise and dimension reduction. Based on Equation 4.7, we can estimate the noise introduced by partitions and control it from the feedback of quality of service (QoS).

Quality-of-service of Partitioned Data

Thanks to Huang et al. [2004], Equation 4.4 is proved to solve in the complexity of polynomial. Once the sparse representation is found, we can either do classification or recovery of original data Y using

$$\text{decide/recover } Y \mapsto \arg\min\ r_i(Y)$$
$$r_i(Y) = \|RY - RD\delta_i(\hat{X})\|_2 + Er_{NVM} + \|\epsilon\|, \tag{4.8}$$

where $(\hat{X})$ is the optimized solution from Equation 4.4 and $\delta_i(\hat{X})$ function is defined as each class, δ_i will selects the sparse vector associated with i class in $\hat{X}$. The reason for robust dense bounded error is that bounded error does not destroy the most correlated features, therefore, the classification or recovery result will not be affected.

As such, the quality of service (QoS) of application is defined as the probability to correctly recognize the class of test data Y by:

$$QoS = 1 - p(e) = 1 - \sum_{i=1}^{M} p(e_i|i)p(i), \tag{4.9}$$

where $p(e)$ refers to the probability of error in classification, and M is the number of classes in data Y. Note that given D and Y, the QoS is mainly affected by the dimension reduction and the resulting $Er_{RAM} + \epsilon$ as we can see from Equation 4.6. As such, given the QoS level, we can have a tolerable maximum error to obtain rp. To make variables clear, definitions of variables are listed in Table 4.4.

STT-MRAM Device Error Model

Featured with fast access speed, high density, and zero standby power, the emerging non-volatile memories (NVM) at nano-scale have introduced promising future for the new non-volatile computing. Among those nano-scale NVM technologies, spin-transfer torque magnetic tunnel junction- (STT-MTJ) based random-access memory (STT-MRAM) has exhibited best poten-tial due to its fast speed (< 10 ns), high integration density ($6 \sim 8F^2$, where F is the feature size) and virtually unlimited endurance ($> 10^{15}$).

Write operation of STT-MRAM can be achieved by injecting current for a period of time. During the write operation, the resistance of STT-MRAM can be changed to high impedance (the 0) or low impedance (the 1). Therefore, one can read the 1-bit information from STT-MRAM by detecting its resistance.

However, the write current of STT-MRAM is usually high to incur large power consump-tion, which cannot satisfy the requirement of low-power applications such as mobile platform

Table 4.4: Variable definitions in sparse-representation-based application

Variables	Definitions
$Y \in \mathbb{R}^d$	Test data
$D \in \mathbb{R}^{d \times t}$	Training data
$X \in \mathbb{R}^t$	Sparse representation of Y
$R \in \mathbb{R}^{d \times t}$	Random projection matrix
E_{RAM}	Error occurred in memory
QoS	Classification successful rate
QoS_c	Constrain of classification successful rate
P_{RAM}	Memory power consumption
P_{max}	Maximum memory power consumption
γ	Dimension reduction ratio
I_0	Critical current of STT-MRAM
I_s	Safe write current of STT-MRAM
I	Applied aggressive scaling current of STT-MRAM
N	Maximum bit width
ω	LSB width

design. As a result, aggressive current scaling can be applied to lower the write energy of STT-MRAM but with write error. The relation between write energy and write current is shown in Equation 4.10:

$$E_{write}(I) = \int_0^T I^2 R(t)dt, \qquad (4.10)$$

where I is the write current applied in STT-MRAM, $R(t)$ the real-time resistance of STT-MTJ and T indicates the write period.

Compared to voltage scaling applied in conventional volatile memories such as SRAM, current scaling in STT-MRAM has the following advantages. For conventional volatile memory, voltage scaling can cause three types of errors: write error, read error, and hold error. In contrast for STT-MRAM, there is no hold error as it does not require to be powered for holding data; and the read current is usually far smaller than write current, so the current scaling will not apply for read operation. The write operation of non-volatile memories, on the other hand, requires a large current that produces enough energy to reverse the magnetization in target cell.

In fact, the switching probability of STT-MRAM cell depends on both applied current amplitude as well as pulse duration. As indicated in Faber et al. [2009], with a fixed pulse width, the switching probability of STT-MRAM cell under different writing current follows

$$p(I) = 1 - \exp(-\frac{I - \alpha}{I_0}),$$ (4.11)

where I_0 is the critical current and α is the fitting parameter in STT-MTJ device model. Critical current I_0 means that switching probability is 0 if $I < I_0$. The relation between switching probability and write current of STT-MRAM under different device parameters is shown in Figure 4.8.

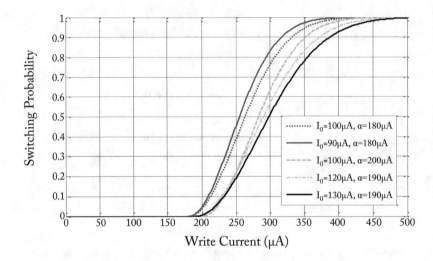

Figure 4.8: The switching probability vs. write current of STT-MRAM under different device parameter.

When a bit fails to switch, the written data will deviate from desired value. Assume the target binary data to write is Y, and the actual data written is $\hat{Y}$ when the error occurs in the corresponding bit, the relation can be formulated as

$$\hat{Y} = Y + \begin{cases} e^+, & Y = 0 \\ e^-, & Y = 1. \end{cases}$$ (4.12)

When the target bit is 0 then the error can only occur when 1 is mis-written, which introduces positive error e^+; and when the target bit is 1 then a negative error e^- will be incurred if writing 1 is unsuccessful. The positive and negative errors subject to the following probability distribution,

$$e^\pm \sim P(e^\pm|I) = \begin{cases} 1 - p(I), & e^\pm = \pm 1 \\ p(I), & e^\pm = 0. \end{cases}$$ (4.13)

Sparse-represented Data Partition on STT-MRAM

As discussed in Section 4.2.1, data partition can be applied to sparse-represented data to save power in storage.

To address data partition in STT-MRAM, critical data called most significant bit (MSB) can be written into STT-MRAM under safety write current named Is to ensure the switching probability, while non-critical data called least significant bit (LSB) can be written under aggressive scaling current I_a, where $I_0 < I_a < Is$. While the errors in LSB may bring small deviation of original data, those in MSB will bring significant impact which is considered unacceptable.

Data partition of sparse-represented data on STT-MRAM is shown in Figure 4.9. The STT-MRAM array is divided in MSB data region and LSB data region. Assuming data is with N bits, we define ω as the LSB width and the non-critical ω bits will be written by I_a while the first critical $(N - \omega)$ bits will be written by I_s. In addition, the boundary between MSB and LSB is flexible in the proposed data partition. An adaptive controller alters LSB width and aggressive scaling current based on the parameter configuration sent by ARM core.

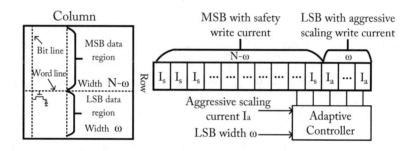

Figure 4.9: Data partition of sparse-represented data on STT-MRAM.

For data partition shown in Figure 4.9, we can obtain the error in system level according to the single bit error model in Equation 4.13, as

$$Er_{NVM} = \sum_{x=1}^{\omega} e_x^{\pm} \times (1 - p(I))$$

$$e_x^{\pm} \sim P(e_x^{\pm}|I) = \begin{cases} 1 - p(I), & e^{\pm} = \pm 2^x \\ p(I), & e^{\pm} = 0, \end{cases} \tag{4.14}$$

where Er_{NVM} is the error occurred by data partition in STT-MRAM in Equation 4.7.

In power consumption perspective, random projection of sparse-represented data can reduce the data volume to store in STT-MRAM according to Equation 4.5. Dimension of data reduces from d to k where $k << d$. We can derive the energy consumption in system level by extending the power model with data partition and without data partition in Equation 4.10:

$$E_{proposed} = \mathbf{N(k)}(\omega E_{write}(I_a) + (N - \omega) E_{write}(I_s))$$
$$E_{original} = \mathbf{N(d)} N E_{write}(I_s),$$

(4.15)

where $\mathbf{N(k)}$ and $\mathbf{N(d)}$ reflect the data volume to store.

According to Equation 4.14 and Equation 4.15, the proposed data partition strategy can save much power with only small compromise on error.

System Architecture

Based on the above-mentioned techniques, we propose an in-memory architecture for adaptive partition and reduction save power without slightly effect on QoS, as shown in Figure 4.10. In this architecture, random projection is applied first for image data from ARM cores to obtain corresponding data in low dimension, which will be stored in STT-MRAM. STT-MRAM is divided into two regions with high write current for MSB and low write current for LSB, respectively. Safe write current I_s will be applied in high write current region while aggressive scaling current I is used in low write current region. After receiving QoS result from OMP, ARM core will alter the parameters in memory by projection controller, LSB width controller and current controller as feedback according to QoS and power consumption shown in Figure 4.10. The proposed architecture features the following designs to overcome aforementioned challenges.

- The non-volatile STT-MRAM is used to keep leakage power minimal.

- The STT-MRAM is equipped with in-memory logic as interface between memory and external processor.

- An in-memory matrix multiplier is designed to perform random projection, which is used to reduce the dimensionality and extract features of both the database image and incoming image. Therefore, the solution to Equation 4.4 is performed under low dimensions to reduce the complexity.

- The STT-MRAM is operated in low current mode by aggressive current scaling to reduce dynamic power. The aggressive current scaling shifts the safe working point of STT-MRAM to a point that has a lower writing current but with slight probability of incurring errors. The errors can be tolerated by the feature based sparse-representation.

- An in-memory OMP logic is deployed to respond to recognition request from external processor by solving Equation 4.4.

- The sparse-representation data is stored in a sparse fashion with index and values.

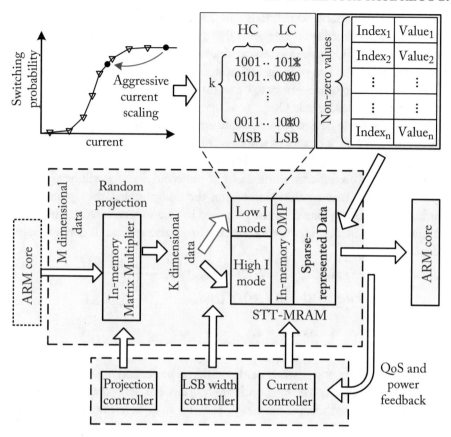

Figure 4.10: The system overview of the proposed sparse-representation data-based architecture.

4.2.2   QOS-AWARE ADAPTIVE CURRENT SCALING

In this section, we first discuss the problem formulation for energy optimization following OMP to solve L_0-norm problem to calculate QoS. Then we use greedy search to find optimization of energy based on look-up table.

Problem Formulation

As the dynamic write energy is the main concern for STT-MRAM, the goal is to optimize the dynamic write energy while maintain the quality of service (QoS). As the proposed architecture aims sparse-representation data, depending on the application. QoS is defined as Equation 4.8.

The three parameters to consider are the scaled current level I, LSB width w, and the dimension reduction ratio γ. The parameter w depicts the width of LSB which is operated in low current mode. The I determines the energy consumed per write operation, w and γ decide the number of necessary operations. Operations with lower current expect to have more errors due to higher probability of unsuccessful magnetization switches as described in Section 4.2.1, which will lead to degradation of QoS. An aggressive dimension reduction, i.e., low γ, will preserve less features and to an extent the loss of information will cause difficulties for recognition thus lower the QoS. The increase of w causes errors to occur on bits with higher significance thus will spoil the feature information and degrade QoS.

In addition, the three design parameters can influence each other. The increase of w will contract the decrease of I as both actions will cause more errors, and which action to take depends on the energy saving margin $\frac{w+1}{w}(\frac{I}{I-\Delta I})^2$. The change of γ also depends on the value of w and I, as the errors incurred on different dimensions of projected features will have different impact on the QoS. To find the best combination within the design space, the optimization problem above is formulated as follow, Based on the proposed in-memory architecture in Figure 4.10, an energy optimization problem under QoS constrain for sparse representation application can be formulated as:

$$\arg\min \ P_{RAM}(\gamma, I, \omega) = P_{\max}\gamma\frac{(N-\omega)I_s^2 + \omega I^2}{NI^2}$$

$$\text{subject to } QoS(\gamma, I, \omega) \geq QoS_c$$
$$0 \leq \omega \leq N$$
$$0 < \gamma \leq 1$$
$$I \leq I_s,$$

(4.16)

where $P_{\max}$ is the maximum power when neither random projection nor data partition is applied, and QoS_c is the QoS constrain given by specific application. For data partition, N represents maximum LSB width. There are three parameters γ, I, and ω to be determined from the architecture to save power.

Consider that QoS is obtained by statistically testing dozens of images by solving Equation 4.4, to find optimized power in the constructed three-dimensional design space is extremely time consuming. In this work, the design space exploration is done in two steps. Firstly, we have examined the relations among all three parameters and their impact on QoS with large dynamic ranges. The core sub-space with largest descent is then identified, in which the optimal solution must exist. Secondly, within the confined sub-space a greedy coordinate searching method is deployed to find the close-optimal solution with small dynamic range.

OMP for STT-MRAM based QoS

QoS is a statistical result from training. Therefore, an efficient solution for Equation 4.4 is necessary. As discussed in Section 4.2.1, L_0-norm solution can be applied to solve Equation 4.4. In this work, we deploy the Orthogonal Matching Pursuit (OMP) solver for the L_0-norm solution, which is a heuristic solver based on greedy algorithm to find the most sparse solution [Tropp and Gilbert, 2007]. The residual is initialized as Y. Index set $\Lambda_0 = 0$ and chosen matrix Φ_0 are empty. The largest correlated column is found by column searching, while index and chosen column will be updated in the active set. The new estimated signal is reconstructed via L_2-norm minimization. The residual is updated from the estimated signal and original signal. In summary, OMP performs two functions as follows. First, it finds the most correlated column from the sensing matrix by comparing simple dot multiplication. Second, the largest correlated column is added to the selected column and by solving an L_2-norm minimization, the most fitted new signal is generated. This procedure will repeat k times to find the expected signal. By using square-root-free Cholesky factorization, the complexity can be reduced from $O(MK^3)$ to $O(MK^2)$, where K is the sparsity. OMP is used to calculate the QoS of the required application and save in a look-up table during the training process. The look-up table is designed as below.

Actions			States	
RP γ	*LSB width ω*	*Current I*	*QoS*	*Energy*
γ_1	ω_1	I_1	QoS_1	E_1
.	.	.	.	.
.	.	.	.	.
.	.	.	.	.

Greedy Optimization

The ***greedy coordinate*** searching method is applied to locally optimize the three-dimensional subspace as shown in Figure 4.11. Since the system with these three parameters is a non-linear system, solving directly becomes a non-convex problem. As such, numerous time may be consumed while the parameter selection is not the best. What is more, obtaining QoS by hardware can be much less time-consuming. During the parameter optimization, only parameter changes happen in specific hardware and other procedure is done in processor. General processor will calculate the parameters to configure and send to parameter control block.

As there are three parameters in the design space, a three-dimensional coordinate can be constructed with I, γ, and w as axes. The searching method starts from empirical given point, and searches in an iterative way. Within each round, the point has three directions to move, and the one yields highest power saving while meets the QoS requirement will be favored as move decision. The iterations continue until none direction can provide a further reduction of power under the given QoS constraint. Then the optimized parameters will be stored in the look-up table. The state is defined as QoS and power, where by taking actions such changing the writing current, LSB width or random projection will lead us to different states.

4.2.3   STT-RAM BASED HARDWARE MAPPING

In this section, we will introduce the hardware implementation of sparse-representation data based architecture shown in Figure 4.10. In such architecture, dimension reduction ratio γ, aggressive scaling current I_a and LSB width ω are alternative.

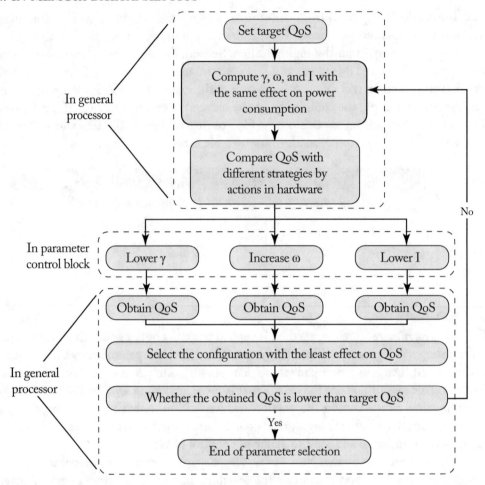

Figure 4.11: Flow of greedy coordinate searching method to optimize three parameters.

Hardware Mapping of Adaptive Random Projection

As discussed in Section 4.2.1, random projection for sparse-representation can be achieved by dimension reduction based on matrix-vector multiplication. Both test data Y and traning data D need to be multiplied by the random projection matrix R. Conventionally, the size of Y, D, and R is fixed if such matrix-vector multiplication can be simply mapped into CMOS-ASIC based implementation. However for the architecture shown in Figure 4.10, dimension reduction ratio γ can be adapted based on greedy coordinate searching method in Figure 4.11. In addition, although the range of reduced dimension k is wide with $0 \le k \le d$ but $k << d$ is the typical case. As a result, redundant energy consumption will occurred if a full-size matrix-vector multiplier is designed.

The adaptive matrix-vector multiplier for proposed architecture is shown in Figure 4.12. Here d_{ij}, r_{ij}, and d'_{ij} represent data in D, R, and D' mentioned in Equation 4.5. Note that $d'_{ij} = \sum_{x=1}^{d} r_{ix} d_{xj}$, N-bit multiplier and adder is required for computation. In the proposed matrix-vector multiplier, a row of result data in D' can be derived simultaneously in $d + 1$ cycles. To calculate the ith row in target matrix, ARM core provides elements in ith row sequently for all multiplier and adder modules. As all the test data and traning data will be multiplied by the same matrix R, each row in R is stored inside a multiplier and adder module to save the bandwidth between ARM core and memory.

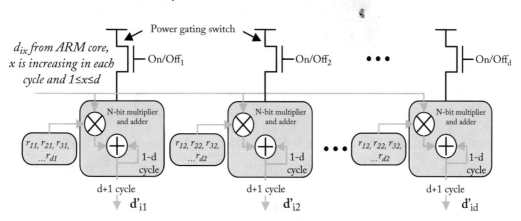

Figure 4.12: Adaptive matrix-vector multiplier for proposed architecture.

To avoid redundant energy consumption, power gating switch is applied to keep the last $d - k$ multiplier and adder modules in sleep mode. ARM core computes corresponding k according to γ and changes power gating status by projection controller.

Hardware Mapping of Adaptive Data Partition
As mentioned in Section 4.2.1, sparse-represented data is stored in STT-MRAM array with flexible LSB width. The LSB controller for data partition in STT-MRAM is shown in Figure 4.13. MSB region and LSB region are the same in physical except the write current is different, which is controlled by multiplexer (MUX). MUX determines the write current for each STT-MTJ by applying V_{MSB} or V_{LSB}, which are the corresponding write voltage of I_s and I_a, respectively.

V_{MSB} is fixed because the safety write current does not change in proposed data partition. To implement adaptive aggressive scaling current I_s, current controller is designed as a multi-level voltage generator as shown in Figure 4.14. I_{ref} is the reference current generated by a fixed bias voltage V_{bias}, and other branch currents are the multiple folder of I_{ref} by current mirror. V_{LSB} can be derived from $V_{LSB} = R I_{sum}$ where I_{sum} is the total current through the resistor. $V_{control}$ is the voltage to switch on/off PMOS to control whether the corresponding current is through the

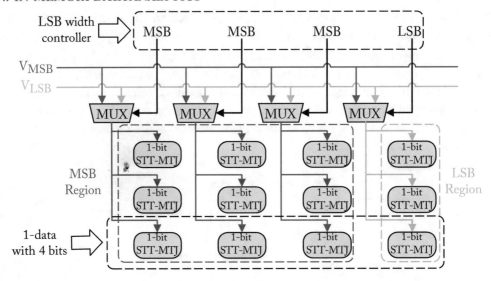

Figure 4.13: Adaptive LSB controller for data partition.

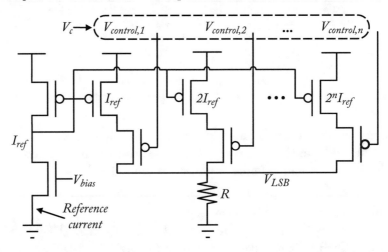

Figure 4.14: Adaptive current controller for data partition.

resistor and V_c is the combination of all $V_{control}$. Assuming n branch current is applied in current controller, it can provide V_{LSB} up to $(2^{n+1} - 1)I_{ref}R$ with high accuracy of $I_{ref}R$.

The relation between V_{LSB}, I_a, and $V_{control}$ can be stored as a look-up table as Table 4.5 by sufficient circuit testing.

Table 4.5: Look-up table of current controller

V_c	V_{LSB}	I_a
$V_{c,1}$	$V_{LSB,1}$	$I_{a,1}$
.	.	.
.	.	.
.	.	.

4.2.4 PERFORMANCE EVALUATION

Design Space Exploration

In this part, we will investigate the relations between QoS and aggressive current scaling parameters, i.e., the least significant bits (LSB) width w and writing current I. To characterize the switching errors of STT-MRAM, the model parameters are set with $I_o = 100$ and $\alpha = 200$, which is fitted with measurement data in Faber et al. [2009] . The switching probabilities at various writing current levels is shown in Table 4.6.

Table 4.6: The STT-MRAM cell switching probability under different I

Current (μA)	550	500	450	400
P_{switch}	1.0000	0.9999	0.9981	0.9817
Current (μA)	350	300	250	200
P_{switch}	0.8946	0.6321	0.2212	0.0000

Under random projection without dimension reduction ($\gamma = 1$), the QoS at different-scaled current levels with varying LSB width w are presented in Figure 4.15a. As indicated in Table 4.6, the error probability increases as the current is scaled down. Therefore, the QoS will decrease as more errors are incurred for low current cases. This is not obvious for a small LSB width w, but significant for large w. Although more energy saving can be anticipated, a large number of w means not only more bits may suffer from errors, but also the magnitude of the errors will increase exponentially. For example, with $w = 1$ the current can be half scaled while the QoS can be almost preserved even approximate one third of the projected features have suffered errors of ± 1 on the last bit. For $w = 6$ at $I \leq 350$, more than one tenth of the projected features will incur errors ranging from minimal ± 1 to maximal $\pm 2^6 - 1$. The deviation is substantial to damage the features, and result shows no successful identification is achieved.

For random projection with $\gamma = 0.8$, the QoS against current results are illustrated in Figure 4.15c. The system with $\gamma = 0.8$ shows better performance compared to their counterparts in Figure 4.15a. For example, in system with $w = 6, \gamma = 0.8$ the current can be half scaled and still provide ideal QoS, yet in system with $w = 6, \gamma = 1$ the current can only be scaled from 600

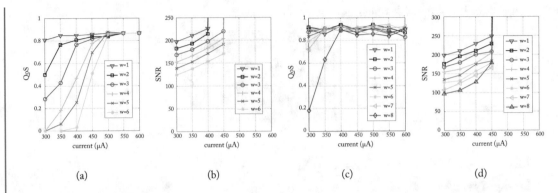

Figure 4.15: Without dimension reduction ($\gamma = 100\%$): (a) the face recognition rate as QoS under different scaled STT-MRAM writing currents with corresponding, (b) signal-to-noise ratio (SNR); with dimension reduction ratio $\gamma = 80\%$, (c) the QoS and (d) signal-to-noise ratio (SNR) under different scaled currents. In all figures, the w denotes the width of LSB that aggressive current scaling is applied to.

to 500. The system performance under $w = 8, \gamma = 0.8$ is comparable to that of $w = 2$ without dimension reduction.

As discussed previously, with dimension reduction the system becomes more robust against errors. This is because of two reasons: (a) projected low dimensional data has expectation of less error and (b) the projected features in low dimension has less fluctuation. As all features are extracted by random projection from the whole image, larger fluctuation means more contradictions among the features, which leads to a higher probability of misclassification in the later recognition stage.

To quantitate how the features are preserved, the signal-to-noise ratio (SNR) is introduced, which is defined as $SNR = 20 \log(\frac{\mu_y}{\sigma_{y-\tilde{y}}})$. The μ_y is the average value of feature vector y, and $\sigma_{y-\tilde{y}}$ is the standard deviation of the introduced errors. The SNR that accords with Figure 4.15a and Figure 4.15c are shown in Figure 4.15b and Figure 4.15d, respectively. For the points with high I, there is no error incurred so the SNR is infinite. From the rest, it can be observed that both the increase of w and decrease of I will cause linear reduction of SNR. The slopes for both linear relations are comparable. In other words, $50\mu A$ reduction of I and an increase of w will cause similar change on the SNR and QoS. As the power saving is proportional to $\frac{w+1}{w}(\frac{I}{I-\Delta I})^2$, we find that in practice $w = 7 \sim 9$ is a roughly optimal configuration.

Figure 4.16a demonstrates the QoS at different dimension reduction ratio. It can be observed that the QoS peaked while γ is around 0.5. For high γ end, the QoS is susceptible to errors, as the features are poorly preserved when errors are incurred. At the low γ end, the QoS damps in a slower slope than that in the high γ end. It can be inferred that the degradation is caused by over-reduction of the dimension instead of errors, as the QoS shows very low dependency on I

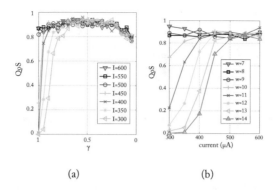

Figure 4.16: (a) Random projection with different dimension reduction rate γ at $w = 7$ and (b) low γ end ($\gamma = 0.2$) tolerance of different w.

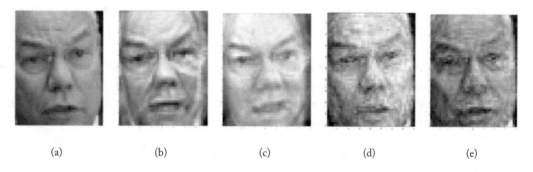

| (a) | (b) | (c) | (d) | (e) |

Figure 4.17: Recovery from down sampling without partitions: (a) original 50×40, (b) dimension reduction 50%, (c) dimension reduction 55%, (d) dimension reduction 60%, and (e) dimension reduction 65%.

and errors. In other words, random projection with low γ is more robust to errors. To verify this, Figure 4.16b shows the QoS under γ with increased w, and it can be observed that while random projection high γ can only endure $w \leq 7$, low γ end is robust to errors with $w \leq 11$.

Figure 4.17 shows that the reconstruction image from dimension reduction. Although from the application points of view, image reconstructions are not necessary. However, as a low power data storage technique, sparse-representation data-based data partition can still recover to the original image with tolerable loss depending on the requirement. Figure 4.17a is the original image and Figure 4.17b–e are different recovered images from dimension reductions. Please note that the original image is only 40x50 size and we are able to recover it from only 1000 samples. Figure 4.18 shows another result from memory partitions. The reconstructed image is less clear than Figure 4.17, however it is still sufficient enough to recognize the person for sparse-representation based applications.

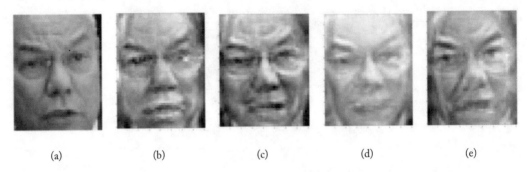

(a)	(b)	(c)	(d)	(e)

Figure 4.18: Recovery from down sampling after partitions 5 bits: (a) original 50×40, (b) dimension reduction 50%, (c) dimension reduction 55%, (d) dimension reduction 60%, and (e) dimension reduction 65% .

The design space is searched by the greedy algorithm as described in Section 4.2.2. The searching step is 0.05 for γ and $50\mu A$ for I. Table 4.7 shows the found parameters for different QoS constraints. The $I^2\gamma$ product is normalized to show the trade-off between QoS constraint and power consumption. The system targeting at 0.95 QoS will consume 18x more write energy compared to the system aims at 0.8. In other words, higher QoS requirement will degrade the system energy efficiency. The parameters $\gamma = 0.1$ and $I = 450\mu A$ with targeted QoS of 0.9 are selected in the following system performance evaluation.

Table 4.7: Design space exploration under different QoS constraints

QoS constraint	γ	I	w	$I^2\gamma$ product
0.95	0.4	450	8	18
0.9	0.1	450	9	4.5
0.85	0.1	300	9	2
0.80	0.05	300	9	1

Accelerator Performance Comparisons

In this part, we compare the proposed accelerator architecture with the baseline system. The baseline system is a typical SRAM for storage and general purpose processor for computation. The comparisons of memory part and logic part are performed separately. To fit the LFW face database size, the memory size is configured to be 16 MB. The leakage power P_l, normalized writing energy $\tilde{E}_w$, and design area is selected as metrics for memory design. For the logic performance comparison, the static power P_s, dynamic power P_d, and design area are used as performance metrics.

For the proposed memory design, the parameters $w = 7$, $I = 450$, and $\gamma = 0.1$ are used, and the memory is configured with four different schemes: (1) non-volatile STT-MRAM only, (2) STT-MRAM with aggressive current scaling, (3) STT-MRAM with random projection, and (4) STT-MRAM with both aggressive current scaling and random projection. The memory performance comparison is listed in Table 4.8. Compared to the SRAM baseline memory, all four schemes with STT-MRAM deployed have reduced the leakage power P_l by 91.4% due to the non-volatility of STT-MRAM. For the dynamic writing operation energy, the STT-MRAM has 3.8x higher writing energy compared to baseline design. This is due to the high current and energy requirement to reverse the magnetization in each cell. With the help of aggressive current scaling, $\tilde{E}_w$ can be improved by 44% for NVM, yet still 2.13x higher than baseline design. Consider the STT-MRAM with random projection deployed, $\tilde{E}_w$ is reduced by 59% compared to the baseline design. And when both techniques are applied simultaneously, the $\tilde{E}_w$ is only one-fifth of the baseline. For memory area, all four schemes are 2.5x times smaller than SRAM. This is because of the cell efficiency of 1T-1MTJ in STT-MRAM compared to the 6T structure of SRAM.

Table 4.8: The performance comparison for memory

Scheme	P_l	$\tilde{E}_w$	Area
Baseline	7.17mW	1.00	86.31 mm^2
NVM		3.80	
NVM + ACS†	614.55μW	2.13	33.83 mm^2
NVM + RP††		0.41	
NVM + ACS + RP		0.21	
† Aggressive current scaling			
†† Random projection			

For the logic comparison, the proposed architecture with additional in-memory OMP solver is compared with general purpose processor only design. The processor model used in Mc-PAT is single-core ARM processor. The OMP solver is implemented by Verilog and synthesized with GlobalFoundries 65 nm low-power PDK. As the random projection can project image data into dimensionality smaller than 100, the OMP solver logic is equipped with the ability to solve 16×16 matrix size maximum. The design constraint is considered for the real-time face recognition in video-level images throughput which exceeds 24 fps per second.

The performance comparison is shown in Table 4.9. The proposed architecture mainly reduces the dynamic power by offloading the computation from ARM core to in-memory logic. Specifically, the dynamic power is reduced by more than three orders of magnitude. On the other hand, the additional in-memory OMP solver has incurred both area and leakage power overhead. Specifically, the area overhead is about 1.4% while the leakage power overhead is negligible.

Table 4.9: The performance comparison for logic

Scheme	P_s	P_d	Area	Cycle	Frequency
Baseline	129mW	232mW	$1.5mm^2$	$10^8 \sim 10^9$	800 MHz
GPP†	129mW	–	$1.5mm^2$	–	–
IML††	31μW	180μW	20942μm^2	~ 160000	3.3 MHz
† General purpose processor					
†† In-memory logic for OMP solver					

4.3   REFERENCES

Candes, E. J., Romberg, J. K., and Tao, T. (2006) Stable signal recovery from incomplete and inaccurate measurements. *Communications on Pure and Applied Mathematics*, 59(8), pp. 1207–1223. DOI: 10.1002/cpa.20124. 123

Dean, J. and Ghemawat, S. (2008) Mapreduce: Simplified data processing on large clusters. *Communications of the ACM*, 51(1), pp. 107–113. DOI: 10.1145/1327452.1327492. 112

Donoho, D. L. (2006) Compressed sensing. *Information Theory, IEEE Transactions on*, 52(4), pp. 1289–1306. DOI: 10.1109/tit.2006.871582. 122

Faber, L. B., et al. (2009) Dynamic compact model of spin-transfer torque based magnetic tunnel junction (mtj). In *Design and Technology of Integrated Systems in Nanoscal Era, DTIS'09. 4th International Conference on, IEEE*, pp. 130–135. DOI: 10.1109/dtis.2009.4938040. 125, 135

Furey, T. S., Cristianini, N., Duffy, N., Bednarski, D. W., Schummer, M., and Haussler, D. (2000) Support vector machine classification and validation of cancer tissue samples using microarray expression data. *Bioinformatics*, 16(10), pp. 906–914. DOI: 10.1093/bioinformatics/16.10.906. 111

Hagan, M. T., Demuth, H. B., Beale, M.H., et al. (1996) *Neural Network Design*. Pws Pub. Boston. 111

Hecht-Nielsen, R. (1994) Context vectors: General purpose approximate meaning representations self-organized from raw data. *Computational Intelligence: Imitating Life*, pp. 43–56. 123

Huang, G. B., Zhu, Q. Y., and Siew, C. K. (2004) Extreme learning machine: A new learning scheme of feedforward neural networks. In *Neural Networks, Proc. IEEE International Joint Conference on*, vol 2, pp. 985–990. DOI: 10.1109/ijcnn.2004.1380068. 111, 124

Huang, G. B., Zhu, Q. Y., and Siew, C. K. (2006) Extreme learning machine: Theory and applications. *Neurocomputing*, 70(1), pp. 489–501. DOI: 10.1016/j.neucom.2005.12.126. 111

Kumar, V., Sharma, R., Uzunlar, E., Zheng, L., Bashirullah, R., Kohl, P., Bakir, M. S., and Naeemi, A. (2014) Airgap interconnects: Modeling, optimization, and benchmarking for back-plane, pcb, and interposer applications. *Components, Packaging and Manufacturing Technology, IEEE Transactions on*, (99), p. 1. DOI: 10.1109/tcpmt.2014.2326798. 119, 120

Liu, S., Pattabiraman, K., Moscibroda, T., and Zorn, B. G. (2012) Flikker: Saving dram refresh-power through critical data partitioning. *ACM SIGPLAN Notices*, 47(4), pp. 213–224. DOI: 10.1145/2248487.1950391. 123

Mairal, J., Bach, F., Ponce, J., and Sapiro, G. (2009) Online dictionary learning for sparse coding. In *Proc. of the 26th Annual International Conference on Machine Learning, ACM*, pp. 689–696. DOI: 10.1145/1553374.1553463. 122

Park, S., Qazi, M., Peh, L. S., and Chandrakasan, A. P. (2013) 40.4 fj/bit/mm low-swing on-chip signaling with self-resetting logic repeaters embedded within a mesh noc in 45 nm soi cmos. In *Proc. of the Conference on Design, Automation and Test in Europe, EDA Consortium*, pp. 1637–1642. DOI: 10.7873/date.2013.332. 119, 120

Parkin, S. S., et al. (2008) Magnetic domain wall racetrack memory. *Science*. DOI: 10.1126/science.1145799. 111

Ranger, C., et al. (2007) Evaluating mapreduce for multi-core and multiprocessor systems. In *Proc. HPCA*. DOI: 10.1109/hpca.2007.346181. 117

Suykens, J. A. and Vandewalle, J. (1999) Least squares support vector machine classifiers. *Neural Processing Letters*, 9(3), pp. 293–300. DOI: 10.1023/A:1018628609742. 111

Thomas, L., Yang, S. H., Ryu, K. S., Hughes, B., Rettner, C., Wang, D. S., Tsai, C. H., Shen, K. H., and Parkin, S. S. (2011) Racetrack memory: A high-performance, low-cost, non-volatile memory based on magnetic domain walls. In *Electron Devices Meeting (IEDM), IEEE International*, pp. 24–2. DOI: 10.1109/iedm.2011.6131603. 111

Tropp, J. A. and Gilbert, A. C. (2007) Signal recovery from random measurements via orthogonal matching pursuit. *IEEE Transactions on Information Theory*, 53(12), pp. 4655–4666. DOI: 10.1109/tit.2007.909108. 130

Venkatesan, R., Kozhikkottu, V., Augustine, C., Raychowdhury, A., Roy, K., and Raghunathan, A. (2012) Tapecache: A high density, energy efficient cache based on domain wall memory. In *Proc. of the 2012 ACM/IEEE International Symposium on Low Power Electronics and Design*, pp. 185–190. DOI: 10.1145/2333660.2333707. 111

Venkatesan, R. K., Herr, S., and Rotenberg, E. (2006) Retention-aware placement in dram (rapid): Software methods for quasi-non-volatile dram. In *High-Performance Computer Architecture. The Twelfth International Symposium on, IEEE*, pp. 155–165. DOI: 10.1109/hpca.2006.1598122. 123

Wang, Y., Yu, H., Ni, L., and Huang, G. B. (2015) An energy-efficient nonvolatile in-memory computing architecture for extreme learning machine by domain wall nanowire devices. *IEEE Transactions on Nanotechnology*, 14(6), pp. 998–1012. DOI: 10.1109/tnano.2015.2447531. 113

Wang, Z., Bovik, A. C., Sheikh, H. R., and Simoncelli, E. P. (2004) Image quality assessment: From error visibility to structural similarity. *Image Processing, IEEE Transactions on* 13(4), pp. 600–612. DOI: 10.1109/tip.2003.819861. 121

Wright, J., et al. (2009) Robust face recognition via sparse representation. *Pattern Analysis and Machine Intelligence, IEEE Transactions on*, 31(2), pp. 210–227. DOI: 10.1109/tpami.2008.79. 123

Yegnanarayana, B. (2004) *Artificial Neural Networks*. PHI Learning Pvt. Ltd. 111

Authors' Biographies

HAO YU

Prof. Hao Yu received Ph.D. degree from the Electrical Engineering Department, University of California, Los Angeles (UCLA) in 2007. He was a Senior Research Staff at Berkeley Design Automation (BDA). Since October 2009, he has been an assistant professor at the School of Electrical and Electronic Engineering, Nanyang Technological University, Singapore. His primary research interest is CMOS emerging technologies at nano-tera scale for energy-efficient data analytics and data links with more than 10M-USD research grant. He has written 200 top-tier peer-reviewed publications, 5 books, and 6 book chapters. Dr. Yu received the Best Paper Award from the ACM Transactions on Design Automation of Electronic Systems (TODAES) in 2010, and Inventor Award from Semiconductor Research Cooperation (SRC) in 2009. He is an associate editor and technical program committee member of many journals and conferences.

LEIBIN NI

Leibin Ni received a B.S. degree in microelectronics from Shanghai Jiao Tong University, Shanghai, China in 2014. He is currently pursuing a Ph.D. degree from the School of Electrical and Electronic Engineering, Nanyang Technological University, Singapore.

His current research interests include emerging non-volatile memory platform and big-data in-memory computing.

YUHAO WANG

Dr. Yuhao Wang received a B.S. degree in microelectronics engineering from Xi'an Jiao Tong University, Xi'an, China in 2011, and a Ph.D. degree in 2015 from the School of Electrical and Electronic Engineering, Nanyang Technological University, Singapore. He is currently a senior R&D engineer at Synopsys, Mountain View, CA, USA.

His research interests include EDA topics related to emerging non-volatile memory design flow and hardware optimization with emphasis on energy efficiency.

Index

Printed in the United States
by Baker & Taylor Publisher Services